Chelates in Nutrition

Authors

F. Howard Kratzer, Ph.D.
Professor Emeritus
Department of Avian Sciences
University of California
Davis, California

Pran Vohra, Ph.D.
Professor
Department of Avian Sciences
University of California
Davis, California

CRC

CRC Press, Inc.
Boca Raton, Florida

Library of Congress Cataloging-in-Publication Data
Main entry under title:

Chelates in nutrition.

Bibliography: p.
Includes index.
1. Minerals in human nutrition. 2. Chelates--physiological effect. 3. Chelation therapy.
I. Kratzer, F. Howard. II. Vohra, Pran.
QP141.C513 1986 612'.392 85-19008
ISBN 0-8493-6544-9

This book represents information obtained from authentic and highly regarded sources. Reprinted material is quoted with permission, and sources are indicated. A wide variety of references are listed. Every reasonable effort has been made to give reliable data and information, but the author and the publisher cannot assume responsibility for the validity of all materials or for the consequences of their use.

Direct all inquiries to CRC Press, Inc., 2000 Corporate Blvd., N.W., Boca Raton, Florida, 33431.

International Standard Book Number 0-8493-6544-9

Library of Congress Card Number 85-19008
Printed in the United States

DEDICATION

We dedicate this book to Georgie Kratzer with sincere thanks for her support, encouragement, and sacrifice of many special projects which were neglected during the preparation of this manuscript.

ACKNOWLEDGMENT

The authors wish to acknowledge the availability of a large number of research reports on chelates and mineral availability which were consulted but not cited due to space considerations.

Every attempt has been made to give credit and obtain permission from the authors and journals for the use of relevant citations, but it is not easy to trace the copyright holders for some old publications. The authors hope to be excused for any such omissions.

AUTHORS

F. Howard Kratzer is Professor Emeritus in the Department of Avian Sciences, University of California, Davis. He received his B.S. degree at Cornell University in 1940 and a Ph. D. degree in nutrition from the University of California, Berkeley in 1944. After 2 years on the faculty of Colorado A & M University at Fort Collins, he joined the University of California at Davis, where he has served as faculty member and a term as Department Chairman until reaching Emeritus status in 1983. He has served on the Editorial Board of the *Journal of Nutrition* and was the Associate Editor for 10 years.

Dr. Kratzer has published over 200 articles covering research on nutritional inhibitors, vitamins, amino acids, and the topic of this book, trace element availability. He is a fellow of the American Association for the Advancement of Science and the Poultry Science Association and a member of many societies, including the American Institute of Nutrition, the American Chemical Society, the Biochemical Society, and the Society for Experimental Biology and Medicine.

Pran Vohra is a Professor in the Department of Avian Sciences, University of California, Davis. He received his B.Sc. (Honour School) and M.Sc. in Chemistry from Panjab University, Lahore in 1940 and 1942, M.S. from Washington State University, Pullman in 1954, and Ph.D. in nutrition from University of California, Davis in 1958. He was on the Editorial Board of Poultry Science and is a member of the Poultry Science Association, American Institute of Nutrition, the Nutrition Society, and the Biochemical Society.

Dr. Vohra has published about 200 articles covering food yeast, biosynthesis of penicillin, energy metabolism, single carbon metabolism, effect of chelates on mineral metabolism, and nutritional evaluation of cereals and legumes.

TABLE OF CONTENTS

Chapter 1

PREFACE AND PERSONAL PERSPECTIVE

Preliminary to a discussion of the function of chelates in nutrition, it may be of interest to consider the series of events which led the authors to recognize that chelating agents may be important in mineral nutrition.

Vitamin B_{12} was discovered in 1948 and the search for other unidentified growth factors has continued ever since. During the 1950s, it was well established that turkey poults grew better when fed a conventional diet containing soybean meal than when fed a purified diet containing isolated soybean protein. This implied that soybean meal must contain some unidentified growth factor for poults. We had been adding Zn at six parts per million to our purified diet for turkey poults for a number of years without realizing that this level was inadequate. In fact, there was a controversy over the need for the addition of Zn to a purified or a practical diet. However, parakeratosis in the rat had been cured by supplementing the diet with Zn during 1941 by Follis et al.[1] and in swine by Tucker and Salmon[2] in 1955. Somehow, the level of Ca in the diet was also important. A higher level of Ca increased the incidence of a parakeratosis-like syndrome in comparison to a lower level.

In 1957, O'Dell and Savage[3] at the University of Missouri showed that additional dietary Zn improved the growth of chicks fed a ration containing isolated soybean protein, and relatively high levels of Ca and P. We also found that poults fed a purified diet containing isolated soybean protein required approximately 40 ppm of supplementary Zn in addition to the 26 ppm present in the basal diet to prevent perosis and give optimum growth. Optimum growth without any perosis was observed without supplementary Zn if isolated soybean protein was replaced with soybean meal.[4] Actually, soybean meal contained 66 ppm Zn compared to 24 ppm in the isolated soybean protein available at the time (Drackett Protein). It was not clear whether the association of the hock disorder with the use of isolated soybean protein was due to its low Zn content or the presence of other factors in this protein that interfered with Zn availability. Supplee et al.[5] made observations on the need of supplementary Zn for poults, but they also had a response with K supplementation. Further studies by many investigators confirmed that the requirement for Zn was greater from diets containing isolated soybean protein than from diets containing other protein sources such as casein or heated, dried egg white.

It seemed logical to test the mineral components of soybean meal for growth and antiperotic activity by ashing soybean meal for a feeding test. However, our attempts to ash the needed amounts of soybean meal in cast-iron crucibles over gas burners in a laboratory hood created such an unusual odor throughout the building that the campus fire department dutifully paid us a call. We did find that the ash was responsible for the improved growth and perosis prevention in poults. This confirmed the role of Zn for the prevention of perosis, but some organic fractions of soybean meal still had growth-promoting properties in the absence of supplementary Zn. Perosis was observed in these cases. Because of the poor availability of Zn from isolated soybean protein, a study of the cause of it for poults was initiated. We suspected that some component of the isolated soybean protein was binding the Zn and making it unavailable. Supplee et al.[5] found that steam-sterilized soybean protein caused less perosis than nonheated protein. We conducted an experiment to compare autoclaved and untreated isolated soybean protein for poults in Zn-deficient purified diets with and without supplementary Zn. If Zn were complexed with a protein-like molecule, autoclaving the isolated soybean protein would denature it and release the Zn. Again, we were not sure whether the Zn of complexes with protein and synthetic complexing agents was effectively utilized. For this reason, a complex of Zn with ethylenediaminetetraacetic acid (EDTA) was also tested. This selection was fortuitous because a colleague was using it in the estimation of Ca. To complete the experimental design, we also included in the study the addition of EDTA alone, without Zn in about the same amount as would be present in the Zn-EDTA complex. We found an improvement in growth and a reduction in the incidence of perosis with autoclaved isolated soybean protein. Zn appeared to be completely available from the EDTA complex and, to our surprise, EDTA by itself gave as good a response as Zn alone in the absence of any supplementary Zn with the untreated protein. Citric acid, when fed at an equimolar level as EDTA, caused no release of Zn from isolated soybean protein. We speculated that EDTA was a stronger chelating agent than the complexing component in the isolated soybean protein and could thus release the Zn from its bound form. The EDTA complex was presumably a molecule small enough for ready absorption through the gastrointestinal (GI) tract. Zn was made readily available to the animal in this way. We also considered the possibility that EDTA was acting by reducing the availability of the Ca, which was known to interfere with Zn utilization, but using a lower Ca level in the diet did not greatly improve the growth of the poults. Later studies indicated that about 100 ppm EDTA alone was as good as 8 ppm of Zn in the diet for growth and prevention of perosis in poults.[7] Our studies were with quite low levels of dietary EDTA.

Several chelating agents were tested for their ability to improve the availability of Zn from a diet containing isolated soybean protein using a biological assay with turkey poults. Of the some 28 chelating agents tested, hydroxyethylethylenediaminetriacetic acid (HEDTA) and ethylenediaminediacetic acid-dipropionic acid (EDDADP) were equal to or slightly better than EDTA, while dihydroxyethylglycine (DHEG), diaminocyclohexanetetraacetic acid (CDTA), and triethylenetetramine (TETA) were essentially ineffective. When the stability constants for the chelating agents as given in the literature were compared with the effectiveness of the chelating agent for improving Zn availability, most of these chelating agents fitted a curve which showed a maximum growth-promoting activity for compounds with stability constants under about 14. Chelating agents, such as diethylenetriaminepentaacetic acid (DTPA) and CDTA, with greater stability constants for Zn than EDTA, had lower growth-promoting activity. The complete lack of activity for CDTA and DTPA was thought to be due to their inability to release the metal for use by the animal after absorption.[8] Unfortunately, this concept had its flaws. Zn is available from complexes with DTPA and CDTA as well as EDTA.[9]

The nature of the complexing agent for Zn in the isolated soybean protein with which the dietary chelate must compete for Zn was an important consideration in this theory of chelate function. O'Dell and Savage[10] in 1960 added a casein-phytic acid complex to a diet for chicks and found a severe growth depression which could be overcome by the addition of Zn to the diet. Since most of the P is present as phytic acid in isolated soybean protein, they suggested that the phytic acid was in some way responsible for reducing the availability of Zn for chickens and turkey poults. Further work showed that an increased Ca level would depress the growth of chickens fed a diet containing phytic acid; this was corrected by the addition of Zn. Ca phytate itself did not depress growth except at rather high levels. On the other hand, phytic acid in combination with casein or glucose, or in a free form, aggravated a deficiency of Zn. This suggested that a Ca-Zn phytate interaction was involved in reducing the biological availability of Zn. They showed no effect of dietary phytic acid on Zn excretion in the urine, but it increased the fecal elimination of Zn about twofold. Phytic acid had no effect on the absorption of dietary proteins.

The binding of Zn by isolated soybean protein was demonstrated by in vitro studies in which a test material was suspended in a solution of radioactive Zn and the supernatant solution counted for activity. The partial removal of the phytic acid from the isolated soybean protein greatly reduced its Zn-binding capacity and conversely, the addition of sodium phytate greatly increased this capacity.[11] EDTA or citric acid effectively removed Zn from the isolated soybean protein and brought it into solution. Both the nutritional and in vitro data indicate that phytic acid in isolated soybean protein was responsible for the poor availability of Zn in diets containing this protein.

The function of a chelate in this system then, appears to be its ability to compete with phytic acid for its Zn-binding capacity and form a soluble complex, but its Zn is available to the animal. It is not known whether the chelate complex itself is absorbed or whether the Zn is released from the complex at the site of absorption. Studies with humans have shown that EDTA is essentially unabsorbed from the intestinal tract. EDTA solutions are injected to detoxify heavy metals in animals and man. However, the dissertation studies of Darwish[12] showed that about 45% of orally administered ^{14}C EDTA was absorbed in the chicken with about 10% excreted in the urine and 5% appearing in the respiratory CO_2 in 48 hr. Much of the activity in the urine was in the form of EDTA, but a portion of the ^{14}C was also found in uric acid, indicating that EDTA was partially metabolized.

Perfusion studies were also carried out in which radioactive complexes of EDTA and Zn or Ca were introduced into isolated segments of the small intestine of chickens and mesenteric blood samples obtained periodically. The parallel appearance of ^{14}C with ^{65}Zn or ^{45}Ca in the mesenteric blood suggested that EDTA and the cations passed through the intestinal mucosa at the same time.[13] It is probable that despite the incomplete metabolism of the chelating agent, enough Zn is released to the animal to meet its need for this nutrient on a Zn-deficient diet.

The absorption of the chelating agent is questioned only when one considers synthetic chelating agents like EDTA. There are many natural materials, such as sugars, proteins, and amino acids, which are strong chelating agents and are certainly biologically available. It is easy to imagine that these dietary components might interact with minerals in the GI tract to form complexes which can be absorbed at faster or slower rates than the free mineral. These naturally occurring chelating agents can be metabolized to release the metal for its specific function in the animal.

In plant nutrition, chelates have been used extensively to improve the availability of certain trace elements.[14] A variety of chelated minerals is commercially available to improve mineral utilization by plants in certain types of soils.

In this book we wish to explore the nature of chelating agents and the reasons for their metal-binding properties. We will then discuss the mechanisms of absorption for various metals and the possible role of chelating agents in influencing the utilization of certain minerals. We will not attempt a complete review of the work in the field, but hope to cite representative data relating to the function of chelating agents in the nutrition of man and animals. We apologize to authors whose good work is not cited in this critical, rather than exhaustive, review.

REFERENCES

1. **Follis, R. H., Day, H. G., and McCollum, E. V.,** Histological studies of the tissues of rats fed a diet extremely low in zinc, *J. Nutr.*, 22, 223, 1941.
2. **Tucker, H. F. and Salmon, W. D.,** Parakeratosis or zinc deficiency in the pig, *Proc. Soc. Exp. Biol. Med.*, 88, 613, 1955.
3. **O'Dell, B. L. and Savage, J. E.,** Potassium, zinc and distiller's dried solubles as supplements to a purified diet, *Poult. Sci.*, 36, 459, 1957.
4. **Kratzer, F. H., Vohra, P., Allred, J. B., and Davis, P. N.,** Effect of zinc upon growth and incidence of perosis in turkey poults, *Proc. Soc. Exp. Biol. Med.*, 98, 205, 1958.

5. **Supplee, W. C., Combs, G. F., and Blamberg, D. L.,** Zinc and potassium effects on bone formation, feathering and growth of poults, *Poult. Sci.*, 37, 63, 1958.
6. **Kratzer, F. H., Allred, J. B., Davis, P. N., Marshall, B. J., and Vohra, P.,** The effect of autoclaving soybean protein and the addition of EDTA on the availability of dietary zinc for poults, *J. Nutr.*, 68, 313, 1959.
7. **Kratzer, F. H. and Starcher, B.,** Quantitative relation of EDTA to availability of zinc for turkey poults, *Proc. Soc. Exp. Biol. Med.*, 113, 424, 1963.
8. **Vohra, P. and Kratzer, F. H.,** Influence of various chelating agents on the availability of zinc, *J. Nutr.*, 82, 249, 1964.
9. **Vohra, P. and Kratzer, F. H.,** Influence of various phosphates and other complexing agents on the availability of zinc for turkey poults, *J. Nutr.*, 89, 106, 1966.
10. **O'Dell, B. L. and Savage, J. E.,** Effect of phytic acid on zinc availability, *Proc. Soc. Exp. Biol. Med.*, 103, 304, 1960.
11. **Allred, J. B., Kratzer, F. H., and Porter, J. W. G.,** Some factors affecting the *in vitro* binding of zinc by isolated soybean protein and by α-casein, *Br. J. Nutr.*, 18, 575, 1964.
12. **Darwish, N. M. and Kratzer, F. H.,** Metabolism of ethylenediaminetetraacetic acid (EDTA) by chickens, *J. Nutr.*, 86, 187, 1965.
13. **Koike, T. L., Vohra, P., and Kratzer, F. H.,** Intestinal absorption of zinc or calcium — EDTA complexes in chickens, *Proc. Soc. Exp. Biol. Med.*, 117, 483, 1964.
14. **Wallace, A.,** Review of chelation in plant nutrition, *Agric. Food Chem.*, 11, 103, 1963.

Chapter 2

CHELATES AND CHELATION

I. METAL COMPLEXES AND CHELATES

Metal complexes are compounds in which the metal atom is linked by more bonds than usually expected by valency considerations. For a long time, metal complexes such as potassium ferrocyanide were known in which two stable molecules, in this case $Fe(CN)_6$ and KCN, were linked together in such a way that no ferrous or cyanide ions were released in solution. Potassium ion was produced, however. The chemical composition of this compound could be represented as $Fe(CN)_6 4KCN$. Werner[1] tried to explain the constitution of these metal complexes in terms of principal and auxiliary valencies. The ionizable atoms or groups were attached in these complexes by principal valencies in an outer sphere of combination and the auxiliary valencies were involved in the coordination of atoms, molecules, and radicals to the central atom in an inner sphere of combination. This is the start of the term coordinate bond. Potassium ferrocyanide is represented as $K_4[Fe(CN)_6]$.

The inert gases show no chemical activity and this is explained in terms of their electronic configuration or distribution of electrons in their various orbitals around the nucleus. According to the classical theory of valency, an atom combined with another atom or entered into a bond in order to achieve an inert gas electronic structure for each of the atoms.

In order of increasing distance, the orbitals of an element are 1s; 2s, 2p; 3s, 3p, 3d; 4s, 4p, 4d, 4f; 5s, 5p, 5d, 5f, 5g; and so on. Each orbital can have a maximum of two electrons of opposite spin.

The bonds may be polar (heteropolar, ionic, electrovalent), nonpolar (homopolar, covalent), and semipolar (coordinate, dative, co-ionic, donor-acceptor) in nature. Coordinate bonds are essentially covalent in nature but the two electrons are originally associated with one atom. Transitional elements in the periodic table (Table 1) are well known for their metal complexes. They tend to attain the nearest inert gas structure by linking with the required number of coordinating groups utilizing all the nine orbitals, namely one s, three p, and five d for bonding or nonbonding electron pairs or for single electrons. The electron configuration for the ground state of elements is given in Table 2.

The term transitional elements was applied to three groups of elements placed in group 8 of the periodic table. The current usage in chelation is much wider and also covers Cu and Zn, Ag and Cd, and Au and Hg, which have many chemical properties similar to those of transition elements.

The metal complexes are coordination compounds in contrast to conventional salts where the metal ion is bound by electrovalency. The primary valency of a metal ion in a salt is the electrovalency, as in NaCl. However, in metal complexes the donor atoms donate electrons to an acceptor metal atom and once this donor-acceptor complex is formed, it is very difficult to distinguish this coordinate bond from a covalent bond. The electron-donating atom is called a ligand atom and the molecule containing the ligand atom is a ligand or ligand molecule. Ligands may be considered as Lewis bases which can share an electron pair with metals. The metals may be regarded as Lewis acids and coordination is a result of neutralization of a Lewis acid with a Lewis base. The resulting complexes may rate from being highly stable to highly ionic. The metal complexes retain their identity to some extent in solution despite their appreciable disassociation. Actually, the term coordination compound is extremely imprecise.

Table 1
PERIODIC TABLE OF THE ELEMENTS

1a	2a	3b	4b	5b	6b	7b	8			1b	2b	3a	4a	5a	6a	7a	0	Orbit
1 H +1 −1 1.0079 1		KEY TO CHART Atomic Number → **50** +2 ← Oxidation States Symbol → Sn +4 Atomic Weight → 118.69 18 18 4 ← Electron Configuration															**2** He 0 4.00260 2	K
3 Li +1 6.94, 2-1	**4** Be +2 9.01218 2-2											**5** B +3 10.81 2-3	**6** C +2 +4 −4 12.011 2-4	**7** N +1 +2 +3 +4 +5 −1 −2 −3 14.0067 2-5	**8** O −2 15.9994 2-6	**9** F −1 18.99840 2-7	**10** Ne 0 20.17, 2-8	K-L
11 Na +1 22.98977 2-8-1	**12** Mg +2 24.305 2-8-2	Transition Elements					Group 8			Transition Elements		**13** Al +3 26.98154 2-8-3	**14** Si +2 +4 −4 28.086 2-8-4	**15** P +3 +5 −3 30.97376 2-8-5	**16** S +4 +6 −2 32.06 2-8-6	**17** Cl +1 +5 +7 −1 35.453 2-8-7	**18** Ar 0 39.948 2-8-8	K-L-M
19 K +1 39.09, -8-8-1	**20** Ca +2 40.08 -8-8-2	**21** Sc +3 44.9559 -8-9-2	**22** Ti +2 +3 +4 47.90 -8-10-2	**23** V +2 +3 +4 +5 50.941, -8-11-2	**24** Cr +2 +3 +6 51.996 -8-13-1	**25** Mn +2 +3 +4 +7 54.9380 -8-13-2	**26** Fe +2 +3 55.847 -8-14-2	**27** Co +2 +3 58.9332 -8-15-2	**28** Ni +2 +3 58.71 -8-16-2	**29** Cu +1 +2 63.546 -8-18-1	**30** Zn +2 65.38 -8-18-2	**31** Ga +3 69.72 -8-18-3	**32** Ge +2 +4 72.59 -8-18-4	**33** As +3 +5 −3 74.9216 -8-18-5	**34** Se +4 +6 −2 78.96 -8-18-6	**35** Br +1 +5 −1 79.904 -8-18-7	**36** Kr 0 83.80 -8-18-8	-L-M-N
37 Rb +1 85.467, -18-8-1	**38** Sr +2 87.62 -18-8-2	**39** Y +3 88.9059 -18-9-2	**40** Zr +4 91.22 -18-10-2	**41** Nb +3 +5 92.9064 -18-12-1	**42** Mo +6 95.94 -18-13-1	**43** Tc +4 +6 +7 98.9062 -18-13-2	**44** Ru +3 101.07 -18-15-1	**45** Rh +3 102.9055 -18-16-1	**46** Pd +2 +4 106.4 -18-18-0	**47** Ag +1 107.868 -18-18-1	**48** Cd +2 112.40 -18-18-2	**49** In +3 114.82 -18-18-3	**50** Sn +2 +4 118.69 -18-18-4	**51** Sb +3 +5 −3 121.75 -18-18-5	**52** Te +4 +6 −2 127.60 -18-18-6	**53** I +1 +5 +7 −1 126.9045 -18-18-7	**54** Xe 0 131.30 -18-18-8	-M-N-O
55 Cs +1 132.9054 -18-8-1	**56** Ba +2 137.34 -18-8-2	**57*** La +3 138.9055 -18-9-2	**72** Hf +4 178.49 -32-10-2	**73** Ta +5 180.947, -32-11-2	**74** W +6 183.85 -32-12-2	**75** Re +4 +6 +7 186.2 -32-13-2	**76** Os +3 +4 190.2 -32-14-2	**77** Ir +3 +4 192.22 -32-15-2	**78** Pt +2 +4 195.09 -32-16-2	**79** Au +1 +3 196.9665 -32-18-1	**80** Hg +1 +2 200.59 -32-18-2	**81** Tl +1 +3 204.37 -32-18-3	**82** Pb +2 +4 207.2 -32-18-4	**83** Bi +3 +5 208.9808 -32-18-5	**84** Po +2 +4 (209) -32-18-6	**85** At (210) -32-18-7	**86** Rn 0 (222) -32-18-8	-N-O-P
87 Fr +1 (223) -18-8-1	**88** Ra +2 226.0254 -18-8-2	**89**** Ac +3 (227) -18-9-2	**104** — +4 (261) -32-10-2	**105** — (262) -32-11-2	106 (263) 32-12-2													O P Q

*Lanthanides	**58** Ce +3 +4 140.12 -20-8-2	**59** Pr +3 140.9077 -21-8-2	**60** Nd +3 144.24 -22-8-2	**61** Pm +3 (145) -23-8-2	**62** Sm +2 +3 150.4 -24-8-2	**63** Eu +2 +3 151.96 -25-8-2	**64** Gd +3 157.25 -25-9-2	**65** Tb +3 158.9254 -27-8-2	**66** Dy +3 162.50 -28-8-2	**67** Ho +3 164.9304 -29-8-2	**68** Er +3 167.26 -30-8-2	**69** Tm +3 168.9342 -31-8-2	**70** Yb +2 +3 173.04 -32-8-2	**71** Lu +3 174.97 -32-9-2	N O P
**Actinides	**90** Th +4 232.0381 -18-10-2	**91** Pa +5 +4 231.0359 -20-9-2	**92** U +3 +4 +5 +6 238.029 -21-9-2	**93** Np +3 +4 +5 +6 237.0482 -22-9-2	**94** Pu +3 +4 +5 +6 (244) -24-8-2	**95** Am +3 +4 +5 +6 (243) -25-8-2	**96** Cm +3 (247) -25-9-2	**97** Bk +3 +4 (247) -27-8-2	**98** Cf +3 (251) -28-8-2	**99** Es +3 (254) -29-8-2	**100** Fm +3 (257) -30-8-2	**101** Md +2 +3 (258) -31-8-2	**102** No +2 +3 (259) -32-8-2	**103** Lr +3 (260) -32-9-2	O P Q

Numbers in parentheses are mass numbers of most stable isotope of that element

Table 2
ELECTRONIC CONFIGURATION OF THE ELEMENTS

Atomic No.	Element	K	L 2		M 3			N 4				O 5				P 6				Q 7			
		1 s	s	p	s	p	d	s	p	d	f	s	p	d	f	s	p	d	f	s	p	d	f
1	H	1																					
2	He	2																					
3	Li	2	1																				
4	Be	2	2																				
5	B	2	2	1																			
6	C	2	2	2																			
7	N	2	2	3																			
8	O	2	2	4																			
9	F	2	2	5																			
10	Ne	2	2	6																			
11	Na	2	2	6	1																		
12	Mg	2	2	6	2																		
13	Al	2	2	6	2	1																	
14	Si	2	2	6	2	2																	
15	P	2	2	6	2	3																	
16	S	2	2	6	2	4																	
17	Cl	2	2	6	2	5																	
18	Ar	2	2	6	2	6																	
19	K	2	2	6	2	6	—	1															
20	Ca	2	2	6	2	6	—	2															
21	Sc	2	2	6	2	6	1	2															
22	Ti	2	2	6	2	6	2	2															
23	V	2	2	6	2	6	3	2															
24	Cr	2	2	6	2	6	5*	1															
25	Mn	2	2	6	2	6	5	2															
26	Fe	2	2	6	2	6	6	2															
27	Co	2	2	6	2	6	7	2															
28	Ni	2	2	6	2	6	8	2															
29	Cu	2	2	6	2	6	10*	1															
30	Zn	2	2	6	2	6	10	2															
31	Ga	2	2	6	2	6	10	2	1														
32	Ge	2	2	6	2	6	10	2	2														
33	As	2	2	6	2	6	10	2	3														
34	Se	2	2	6	2	6	10	2	4														
35	Br	2	2	6	2	6	10	2	5														
36	Kr	2	2	6	2	6	10	2	6														
37	Rb	2	2	6	2	6	10	2	6	—		1											
38	Sr	2	2	6	2	6	10	2	6	—		2											
39	Y	2	2	6	2	6	10	2	6	1		2											
40	Zr	2	2	6	2	6	10	2	6	2	—	2											
41	Nb	2	2	6	2	6	10	2	6	4*	—	1											
42	Mo	2	2	6	2	6	10	2	6	5	—	1											
43	Te	2	2	6	2	6	10	2	6	6	—	1											
44	Ru	2	2	6	2	6	10	2	6	7	—	1											
45	Rh	2	2	6	2	6	10	2	6	8	—	1											
46	Pd	2	2	6	2	6	10	2	6	10*	—	0											
47	Ag	2	2	6	2	6	10	2	6	10	—	1											
48	Cd	2	2	6	2	6	10	2	6	10	—	2											
49	In	2	2	6	2	6	10	2	6	10	—	2	1										
50	Sn	2	2	6	2	6	10	2	6	10	—	2	2										
51	Sb	2	2	6	2	6	10	2	6	10	—	2	3										
52	Te	2	2	6	2	6	10	2	6	10	—	2	4										
53	I	2	2	6	2	6	10	2	6	10	—	2	5										
54	Xe	2	2	6	2	6	10	2	6	10	—	2	6										
55	Cs	2	2	6	2	6	10	2	6	10	—	2	6	—	—	1							

Table 2 (continued)
ELECTRONIC CONFIGURATION OF THE ELEMENTS

Atomic		K	L 2		M 3			N 4				O 5				P 6				Q 7			
No.	Element	1 s	s	p	s	p	d	s	p	d	f	s	p	d	f	s	p	d	f	s	p	d	f
56	Ba	2	2	6	2	6	10	2	6	10	—	2	6	—	—	2							
57	La	2	2	6	2	6	10	2	6	10	—	2	6	1	—	2							
58	Ce	2	2	6	2	6	10	2	6	10	2*	2	6	—	—	2							
59	Pr	2	2	6	2	6	10	2	6	10	3	2	6	—	—	2							
60	Nd	2	2	6	2	6	10	2	6	10	4	2	6	—	—	2							
61	Pm	2	2	6	2	6	10	2	6	10	5	2	6	—	—	2							
62	Sm	2	2	6	2	6	10	2	6	10	6	2	6	—	—	2							
63	Eu	2	2	6	2	6	10	2	6	10	7	2	6	—	—	2							
64	Gd	2	2	6	2	6	10	2	6	10	7	2	6	1	—	2							
65	Tb	2	2	6	2	6	10	2	6	10	9*	2	6	—	—	2							
66	Dy	2	2	6	2	6	10	2	6	10	10	2	6	—	—	2							
67	Ho	2	2	6	2	6	10	2	6	10	11	2	6	—	—	2							
68	Er	2	2	6	2	6	10	2	6	10	12	2	6	—	—	2							
69	Tm	2	2	6	2	6	10	2	6	10	13	2	6	—	—	2							
70	Yb	2	2	6	2	6	10	2	6	10	14	2	6	—	—	2							
71	Lu	2	2	6	2	6	10	2	6	10	14	2	6	1	—	2							
72	Hf	2	2	6	2	6	10	2	6	10	14	2	6	2	—	2							
73	Ta	2	2	6	2	6	10	2	6	10	14	2	6	3	—	2							
74	W	2	2	6	2	6	10	2	6	10	14	2	6	4	—	2							
75	Re	2	2	6	2	6	10	2	6	10	14	2	6	5	—	2							
76	Os	2	2	6	2	6	10	2	6	10	14	2	6	6	—	2							
77	Ir	2	2	6	2	6	10	2	6	10	14	2	6	7	—	2							
78	Pt	2	2	6	2	6	10	2	6	10	14	2	6	9	—	1							
79	Au	2	2	6	2	6	10	2	6	10	14	2	6	10	—	1							
80	Hg	2	2	6	2	6	10	2	6	10	14	2	6	10	—	2							
81	Ti	2	2	6	2	6	10	2	6	10	14	2	6	10	—	2	1						
82	Pb	2	2	6	2	6	10	2	6	10	14	2	6	10	—	2	2						
83	Bi	2	2	6	2	6	10	2	6	10	14	2	6	10	—	2	3						
84	Po	2	2	6	2	6	10	2	6	10	14	2	6	10	—	2	4						
85	At	2	2	6	2	6	10	2	6	10	14	2	6	10	—	2	5						
86	Rn	2	2	6	2	6	10	2	6	10	14	2	6	10	—	2	6						
87	Fr	2	2	6	2	6	10	2	6	10	14	2	6	10	—	2	6	—	—	1			
88	Ra	2	2	6	2	6	10	2	6	10	14	2	6	10	—	2	6	—	—	2			
89	Ac	2	2	6	2	6	10	2	6	10	14	2	6	10	—	2	6	1	—	2			
90	Th	2	2	6	2	6	10	2	6	10	14	2	6	10	—	2	6	2	—	2			
91	Pa	2	2	6	2	6	10	2	6	10	14	2	6	10	2*	2	6	1	—	2			
92	U	2	2	6	2	6	10	2	6	10	14	2	6	10	3	2	6	1	—	2			
93	Np	2	2	6	2	6	10	2	6	10	14	2	6	10	4	2	6	1	—	2			
94	Pu	2	2	6	2	6	10	2	6	10	14	2	6	10	6	2	6	—	—	2			
95	Am	2	2	6	2	6	10	2	6	10	14	2	6	10	7	2	6	—	—	2			
96	Cm	2	2	6	2	6	10	2	6	10	14	2	6	10	7	2	6	1	—	2			
97	Bk	2	2	6	2	6	10	2	6	10	14	2	6	10	9*	2	6	—	—	2			
98	Cf	2	2	6	2	6	10	2	6	10	14	2	6	10	10	2	6	—	—	2			
99	Es	2	2	6	2	6	10	2	6	10	14	2	6	10	11	2	6	—	—	2			
100	Fm	2	2	6	2	6	10	2	6	10	14	2	6	10	12	2	6	—	—	2			
101	Md	2	2	6	2	6	10	2	6	10	14	2	6	10	13	2	6	—	—	2			
102	No	2	2	6	2	6	10	2	6	10	14	2	6	10	14	2	6	—	—	2			
103	Lr	2	2	6	2	6	10	2	6	10	14	2	6	10	14	2	6	1	—	2			
104	—	2	2	6	2	6	10	2	6	10	14	2	6	10	14	2	6	2	—	2			

[a] Note irregularity.

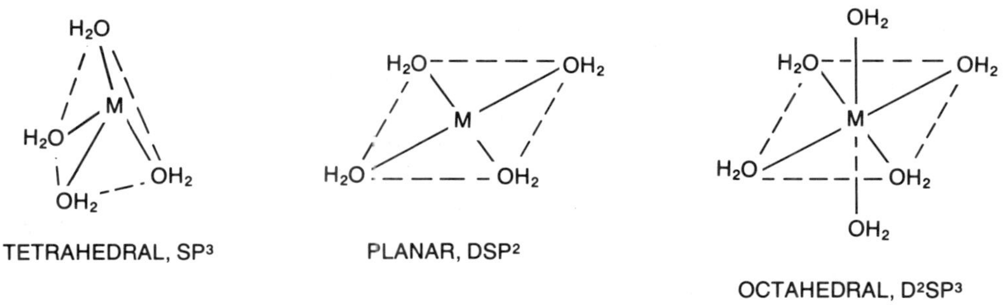

FIGURE 1. Spatial distribution of donor ligand around metal atom.

II. CHELATES

The word Chelate is derived from the Greek word Che'le' for the pincer-like claws of a crab, lobster, or scorpion. Morgan and Drew[2] suggested this term for metal complexes in which the metal atom is held in the complex through more than one point of attachment analogous to a pincer-like grip. Chelation is the process of formation of chelates.

A chelating agent which yields a soluble metal complex is called a sequestering agent. A metal ion may be postulated to exist in a hydratocomplex bound to a number of water molecules in an aqueous solution with the negative oxygen of the water dipole oriented towards the positive metal. This results in a certain spatial distribution or configuration of the donor ligand around the central metal atom. The configuration of these hydratometal complexes may be idealized as seen in Figure 1 and are indicated as sp^3, dsp^2, and d^2sp^3.

Other configurations are also encountered and the bond angles, bond distances, and the attached number of water molecules depend upon the ionic strength, concentration, and temperature.

In general, electrons of the water molecule are more displaced towards the metal ion and hydrogen can be more easily displaced from the bound water than from the unbound water. Other ions or molecules may replace water and the resulting compound is called a metal complex or a metal coordination compound. The bound ligand may be neutral or negatively charged, but electrostatic action and charge redistribution tend towards the metal. The binding of the metal atom to the ligands may be electrostatic or covalent in nature, with electron-pair bonds being most dominant. However, it is the ligand that donates electrons to the metal atom which occupies a central position in the metal complex.

If the ligand has more than one donor atom and the resulting complex has a heterocyclic ring, then it is called a chelate ring and the ligand is a chelating agent. Any metal complex with one or more chelate rings is a chelate of that metal. The ligand atoms are not only attached to the metal ion, but also to each other in metal chelates and the metal is always a part of the heterocyclic ring (Figure 2).

The size or the number of heterocyclic rings in a chelate is not restricted, but the stable chelate rings generally contain five or six atoms. The ring size of some metal-amino acid chelates is given in Table 3.

III. STEREOCHEMISTRY

The chelate rings may be symmetrical or unsymmetrical. The unsymmetrical rings exhibit geometrical (*cis* and *trans*) isomerism (Figure 3). Also, optical isomerism may arise when a compound can be represented by two asymmetrical structures which are mirror images of each other. The example is $[Co(en)_2Cl_2]Cl$ where en is ethylenediamine (Figure 4). Both of these are *cis*-forms and the optically inactive form is the *trans*-form (Figure 5).

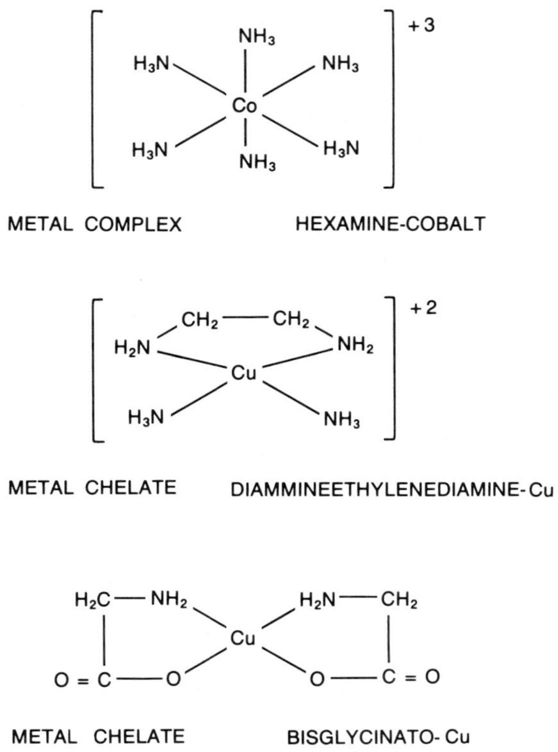

FIGURE 2. Metal complexes and metal chelates.

Table 3
TERDENTATE A-AMINO ACIDS: RING SIZE AND DONOR ATOM

Ring size	Donor atom O	Donor atom N	Donor atom S
5	Serine	2,3-Diaminopropionate	Cysteine
	Threonine		Cystine
			Pencillamine
6	Aspartate	2,4-Diamonbutanoate	Homocysteine
	Asparagine	Asparagine	Methionine
		Histidine	
7	Glutamate	Ornithine	
	Glutamine	Arginine	
8		Lysine	

From Martin, R. B., *Metal Ions in Biological Systems,* Vol. 9, Sigel, H., Ed., Marcel Dekker, New York, 1979. With permission.

Ionization isomers are also encountered depending upon the position of the ion inside or outside the complex. Sulfatopentaminocobaltibromide, $[CO(NH_3)_5SO_4]Br$, gives a precipitate with silver nitrate, but not with barium chloride. Bromopentaminocobaltosulfate, $[Co(NH_3)_5Br]SO_4$, gives a precipitate with barium chloride but not silver nitrate.[4]

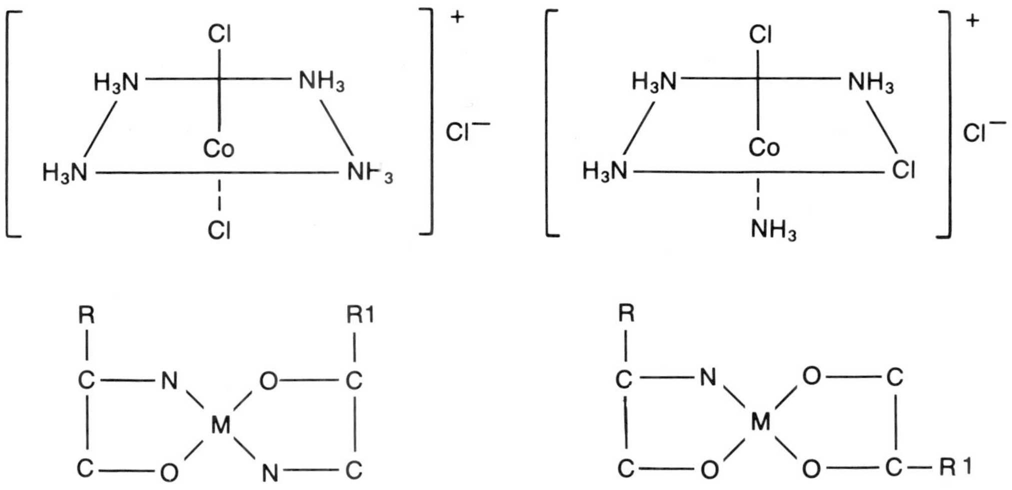

FIGURE 3. Geometrical isomerism in chelates.

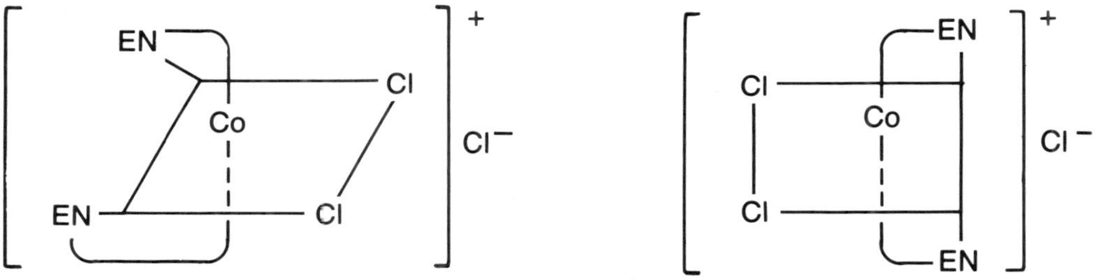

FIGURE 4. Optical isomerism in metal complexes.

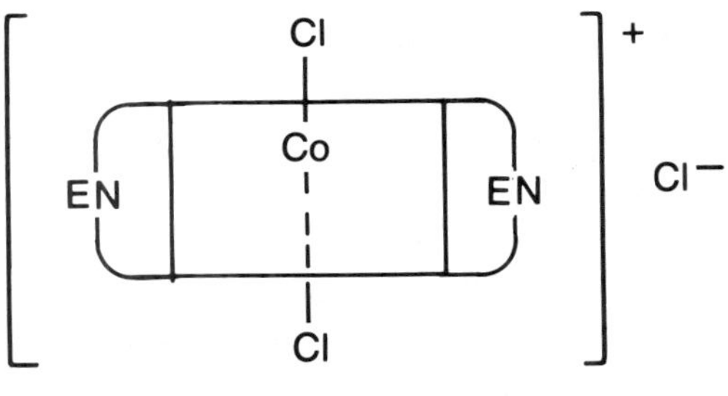

FIGURE 5. Optical inactive *trans* isomer.

IV. NOMENCLATURE

Complex cation — The name of the ligand is followed by the central atom with its oxidation number indicated in parentheses.

$Ag(NH_3)_2^+$ Diamine silver(I)

Complex anion — The name gives the number and names of ligands followed by the name of the central atom with an-ate or -ic ending and its oxidation number in parentheses.

$Zn(OH)_4^=$ Tetrahydroxozincate(II)

Table 4
DIRECTIONAL PROPERTIES OF ORBITALS

Coordination no.	Orbitals used	Configuration	Number: Sides	Number: Corners
2	sp	Linear	—	2
3	sd^2	Trigonal-planar		
4	sp^3	Tetrahedral	4	4
	dsp^2	Square-planar	Flat	4
5	dsp^3	Trigonal-bipyramidal	6	5
6	d^2sp^3	Octahedral	8	6
	d^4sp	Trigonal-prism	5	6
7	d^3sp^3	Face centered octahedral	8	6

Table 5
OXIDATION STATES OF SOME TRANSITIONAL ELEMENTS

Sc (3)
Ti (0,2,3,4)
V (−1,0,1,2,3,4,5)
Cr (−1,0,1,2,3,4,5,6)
Mn (−1,0,1,2,3,4,5,6,7)
Fe (0,2,3,4)
Co (−1,0,1,2,3,4)
Ni (0,1,2,3,4)
Cu (0,1,2,3)

V. COORDINATION NUMBER

The total number of ligand atoms bound directly to the central metal atom gives the coordination number of that metal. It can vary with the nature of the donor atom and the oxidation state of the metal. "In a covalent compound of known structure the oxidation number of each atom is the charge remaining on the atom when each shared electron pair is assigned completely to the more electronegative of the two atoms sharing it. A pair of shared electrons by two atoms of the same element is split between them."[5] Oxidation number is another name for the classical valence of the element. Transition elements exhibit a range of oxidation numbers. A correlation between oxidation number and the coordination number or the maximum number of donor atoms that can be bound simultaneously to the metal is not easily predictable.

In general, a single coordination number is assigned for each metal ion which is satisfied by the attached ligands. The coordination sites have a well-defined stereochemical arrangement in space. However, the coordination number may not have a unique value. For example, pink and blue Co(II) chlorides have coordination numbers of 6 and 4, respectively.

Information about the coordination number, orbitals used, and the spatial arrangement of bonds may be simplified briefly as in Table 4. The oxidation states of some of the transition elements are given in Table 5.

The term transition elements is strictly applicable to the members of the group 8, but extends currently to cover groups 3A, 4A, 5A, 6A, 7A, 8, 1B, and 2B of the periodic table. The members of this series are listed in Table 6. The common coordination requirements of some metal ions and the configurations of their complexes are indicated in Table 7.

Table 6
EXTENDED TRANSITION ELEMENTS

Group										
3A	**4A**	**5A**	**6A**	**7A**	**8**			**1B**	**2B**	**3B**
Sc	Ti	V	Cr	Mn	Fe	Co	Ni	Cu	Zn	Ga
Y	Zr	Nb	Mo	Te	Ru	Rb	Pd	Ag	Cd	In
La	Hf	Ta	W	Re	Os	Ir	Pt	Au	Hg	Tl

Table 7
COORDINATION NUMBER AND CONFIGURATION

Coordination no.	Metal ion
2 (Linear)	Cu(I), Ag(I), Hg(I), Hg(II)
4 (Tetrahedral)	Mg(II), Zn(II), Cd(II), Hg(II)
4 (Planar)	Co(II), Ni(II), Cu(II), Ag(II), Pd(II), Pt(II)
6 (Octahedral)	V(III)(IV), Cr(III), Mn(II)(III), Fe(II)(III), Co(II)(III), Ni(II)(IV), Cd(II), Zn(II), Al(III), Sn(II)(III), Pb(II)(IV), Mo(IV), W(IV)

From Chaberek, S. and Martell, A. E., *Organic Sequestering Agents,* John Wiley & Sons, New York, 1964. With permission.

VI. TYPES OF METAL CHELATES

A chelating agent may attach to the metal through two, three, four, five, and six positions and may be called a bidentate, tridentate, quadridentate, quindentate, or sexadentate chelating agent. A few examples can be seen in Figure 6.

VII. CHELATING AGENTS

A chelating agent must meet the following criteria according to Dweyer and Mellor:[7]

1. It must possess at least two functional groups, each capable of donating a pair of electrons to combine with the metal.

   a. Coordinating groups:

=O	–OR	- - AsR_2
$–NH_2$	–NOH	- - PR_2
–NH	–OH (alcoholic)	
–N=	–S– (thioether)	

   b. After a loss of proton:

–COOH	$–NH_2$
$–SO_3$	–N(R)–H

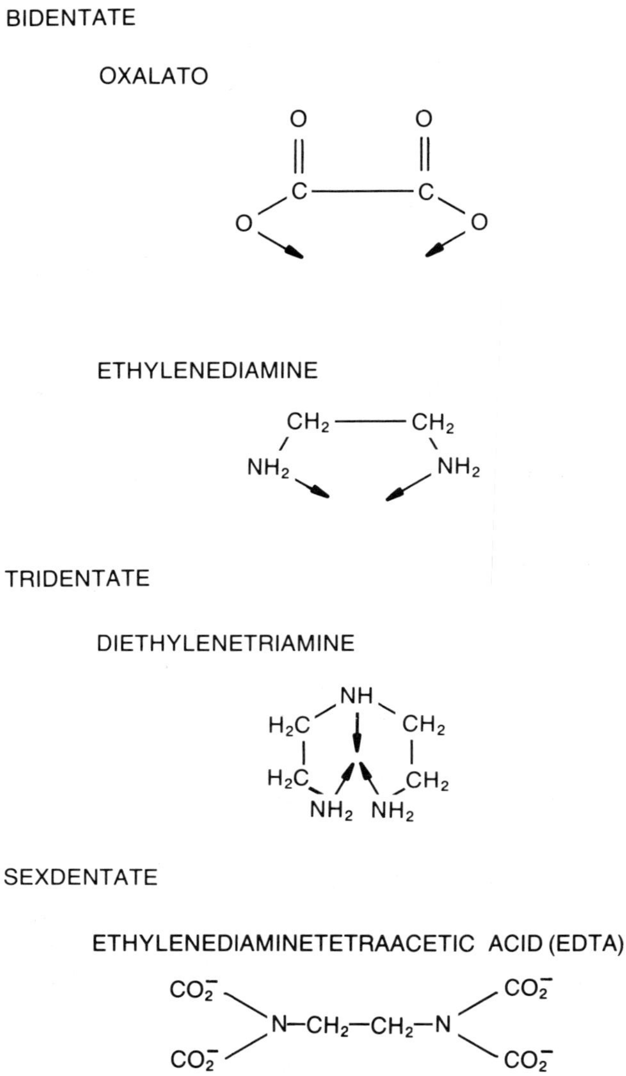

FIGURE 6. Attachment of chelates to metal ions.

–OH (enolic, phenolic)

–N(H)=O

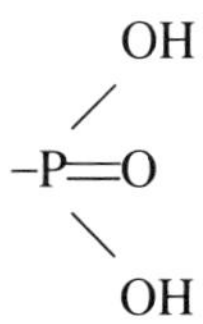

–SH

The donor atoms in chelates are generally:

N	O	Cl
P	S	Br
As	Se	I

The most important chelating agents contain N or O atoms or both. The order of decreasing affinity for the metal atom may be as follows:[8]

$$-O > -NH_2 > -N=N > \text{ring } N > -COO^- > -O- > -C=$$

enolate > amine > azo > ring N > carboxylate > ether > carbonyl

2. The functional groups are so located that a ring structure is formed with the metal atom as the closing member. The potential of ring formation is influenced by stearic factors. Due to stearic hindrances, the attachment of one chelating agent may hinder the attachment of a competing agent.

Besides the synthetic compounds, chelating agents are widely distributed in all living systems in nature. Some of these chelating agents with typical examples include:

Water as a chelating agent:

$$M(H_2O)_x^{2+} + OH^- = M(H_2O)_{x-1}OH^+ + H_2O$$

Carbohydrates and their derivatives:
- Simple sugars: ribose, deoxyribose, glucose, fructose
- Sugar derivatives: glucosamine
- Oligosaccharides: sucrose, lactose, maltose
- Polysaccharides and derivatives: cellulose, modified celluloses, starch, modified starches, pectins, gums, alginic acid, chitin, chitisans
- Acid derivatives of carbohydrates: lactic acid, pyruvic acid, citric acid

Lipids:
- Glycerides mono-, di-, and triglycerides
- Phosphatidic acid derivatives: phosphyatidylcholine, phosphatidylethanolamine, ceramide, sphingomyelin
- Carotenoids: xanthophylls, vitamin A
- Steroids structure: cortisone, cortisole, cholic acid, deoxycholic acid, taurocholic acid

Proteins and their derivatives:
- Amino acids: essential, semiessential, and nonessential

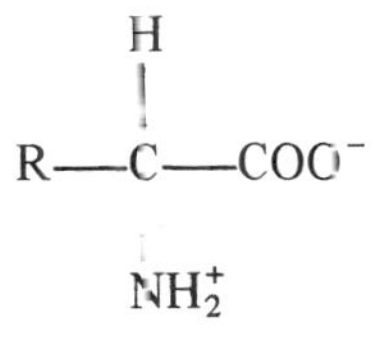

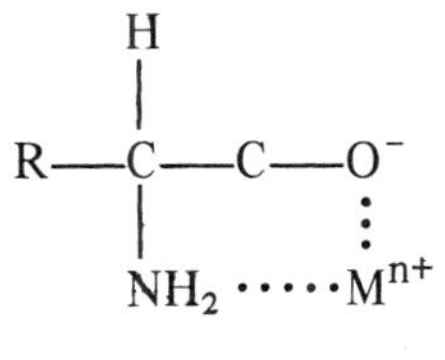

Amino acids: (Hydrophobic R) glycine, alanine, valine, leucine, isoleucine, phenylalanine, (Inert heterotom containing R), tryptophan, (Hydrocyclic R), serine, threonine, tyrosine, (Carboxylic R), aspartic acid, glutamic acid, (Amine R), lysine, arginine, (Amide R), asparagine, glutamine, (Imidazole), histidine, (Sulfyhdryl and thioether), cysteine, methionine, (Other), proline

Amino acid derivatives: histamine, adrenaline

Peptides: protamines

Polypeptides: insulin, growth hormone

Enzymes:

Nucleoproteins:

Nucleotides: ATP, NAD, FMN, FAD, Coenyme A

Nucleic acids: DNA, RNA
Carbonate, bicarbonate
Cyanides
Organic acids with at least one more additional functional group besides the carboxylic: glycolic acid, oxalic acid, lactic acid, citric acid
Phosphates, *ortho,* pyro, *meta,* poly; phytic acid
Phosphonates: MDP (methylenephosphonate), HMDP (hydroxy methylenephosphonate), HEDP (1-hydroxyethylidene) diphosphonate
Poryphyrins: heme-Fe (hemoglobin), chlorophyll
Plant alkaloids
Plant phenolics: anthocyanins, tannins
Sulfates, sulfonic acids, chondroitin sulfates
Vitamins: vitamin B_{12}, ascorbic acid
Some synthetic chelates:
BAL or 2,3-dimercaptopropanol-1
CDTA or Chel 600 (cyclohexane-*trans*-1,2-diaminotetraacetic acid)
DFOA or Desferrioxamine B
DHEEDA or Versenediol (dihyroxyethylethylenediaminediacetic acid)
DHG (*N,N-bis*(2-hydroxyethyl)glycine)
$(HOOCCH_2)_2NCH_2–CH_2–N(CH_2COOH)–CH_2CH_2N(CH_2COOH)_2$
DMPS or sodium 2,3-dimercaptopropane-1-sulfonate
DMSA or 2,3-dimercaptosuccinic acid
DTPA or Chel 330 (diethylenetriaminepentaacetic acid)
EDDHA or EHPG or Chel 138 (ethylenediamine-di(*o*-hydroxy)-phenylacetic acid)
$(HOOCCH_2)_2N–CH_2–CH_2N(CH_2COOH)_2$
EDTA or Versene or Sequestrene AA (ethylenediaminetetraacetic acid),
EGTA or ethyleneglycol-*bis*-(b-aminoethyl ether)*N,N,N′,N′*-tetraacetic acid
EHPG or ((ethylene-*bis-N,N′*-2-*o*-hydroxyphenyl)glycine)
HBEA or (*N,N,N′-tris*(2-hydroxybenzyl)ethylenediamine-*N*-acetic acid)
HBED or (*N,N′-bis.*-(*o*-hydroxybenzyl)ethylenediamine-*N,N*-diacetic acid)
HEDTA or Versenol or HEEDTA or EDTA-OH (*N*-(2-hydroxyethyl)ethylenediamine-*N,N,N′*-triacetic acid)
HIMDA or ((*N*-2-hydroxyethyl)iminodiacetic acid))
Ionophores (Lasalocid, Valinomycin, Enniatin)
NTA or (nitrilotriacetic acid)
$(HOOCCH_2)_3N$
Salicylic acid
TEA or (triethanolamine)
TETHA or TTHA or (triethylenetetraminehexaacetic acid)
THBE (*N,N,N,N′*-tetrakis(2-hydroxygenyl)ethylenediamine)
Unithiole or sodium-2,3-dimercaptopropane-1-sulfonate

VIII. BONDING IN METAL COMPLEXES

A number of theories have evolved to explain the properties of metal complexes. These include the valence bond theory, involving orbitals and their hybridization, and electrostatic crystal field theory, dealing with electrostatic fields imposed by ligands on the different orbitals of the metal. The fields account for light absorption and magnetic properties of the chelates.

The ligand field theory evolved out of the crystal field theory. The occupation of d orbitals follows Hund's rule of maximum multiplicity, which states that the distribution of electrons in a free atom between 3 p, 5 d, and 7 f orbits is such that as many orbits as possible are occupied by single electrons before any pairing of electrons with opposite spin takes place. Suppose six electrons are available for occupying five d orbitals; then one electron with the same spin occupies each of the d orbitals and the sixth electron pairs up with one of the electrons in one of the d orbitals. The chemistry of chelation deals mostly with the interaction of electrons in the five d orbitals with the ligand field.

These theories will not be discussed here in any detail. The nature of bonding in chelates is explained in terms of quantum mechanics utilizing atomic and molecular orbital approximations rather than stationary orbits. The electronic structure of the elements determines their physical and chemical properties. The wave pattern of an electron around the nucleus can be expressed by Schrodinger's intuitive equation where ψ is wave function, m is the mass of the electron, W is the total energy, V is the potential energy, h is Planck's constant, and x, y, and z are the coordinates of the electron:

$$\frac{d^2\Psi}{dx^2} + \frac{d^2\Psi}{dy^2} + \frac{d^2\Psi}{dz^2} + \frac{8\,\pi^2 m}{h^2}(W - V)\Psi = 0$$

This equation can only be solved satisfactorily for certain definite eigenvalues of total energy W corresponding to those of stationary states in Bohr's model of atoms, with whole numbers corresponding to quantum numbers. Wave functions enable one to postulate the shape and other properties of s, p, and d orbitals.

This expression of one-electron function is a product of a function of radial distance of the electron from the nucleus with another function which is dependent only on the angular coordinates of the electron. The angular functions are independent of the element. There is just one angular distribution of s electrons, three for p electrons, and five for d electrons. The spatial properties of these orbitals are represented in Figure 7 in the form of balloons. About 90% of the electron density is included within these shapes.

The relative energies of the electronic orbitals are represented graphically in Figure 8.

The orbitals 1s, 2s, 2p, 3s, and 3p are filled in a successive manner, but the energy level of the 3d orbital is higher than the 4s orbital and the latter becomes filled earlier than the 3d orbitals. This is repeated when 5s fills before 4d, and 6s before 5d. In actual practice, the distinction between s, p, and d orbitals is not very rigid as long as they have comparable energies and these can be combined to form relatively strong hybrid orbitals. In hybridization, the energy level of indicated orbitals becomes equal by assuming an average value.

If the number of orbitals in a cation is insufficient for complexing, the orbitals may assume an excited state to give a new arrangement with sufficient bonding orbitals, as indicated for Co^{3+} and Fe^{3+} (Figure 9).

The distribution of electrons in some elements and ions is shown in Figures 10 and 11.

The five d orbitals in a free ion are termed degenerate and energetically alike, but the electrons occupy these orbitals following Hund's rule of maximum multiplicity.[4] Crystal field theory is concerned with the effect of surrounding ligands on the energy of the d orbitals, which results in resolving or splitting of degeneracy as shown in Figure 12. The phenomenon is termed crystal field splitting.

Let us apply this information in chelation. A metal ion M^{m+} is bound by some X^- anions in an octahedral configuration and has only a single d electron outside the closed shells. This electron could occupy any of the five d orbitals if these were equivalent. The shape of d orbitals in the figure suggests that the three orbitals d_{xy}, d_{yz}, and d_{zx} are equivalent because they form a set called t_{2g} with spherical symmetry directed between the axis and are more favored than orbitals d_{z^2} and $d_{x^2-y^2}$. The latter form another equivalent set called e_g which differs from the first three by forming regions of electron density along the axis. The ligand environment influences these two sets of orbitals differently. The energies of $d_{x^2-y^2}$ or d_{z^2}

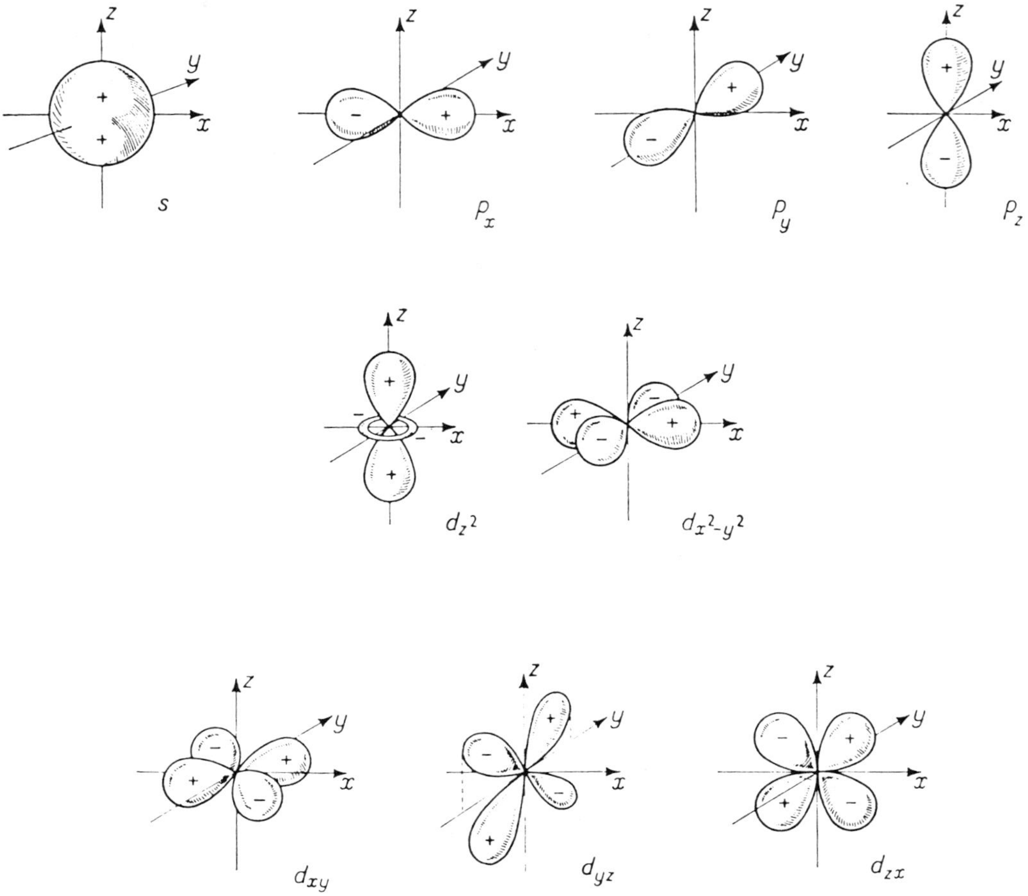

FIGURE 7. Balloon representation of s, p, and d atomic orbitals.

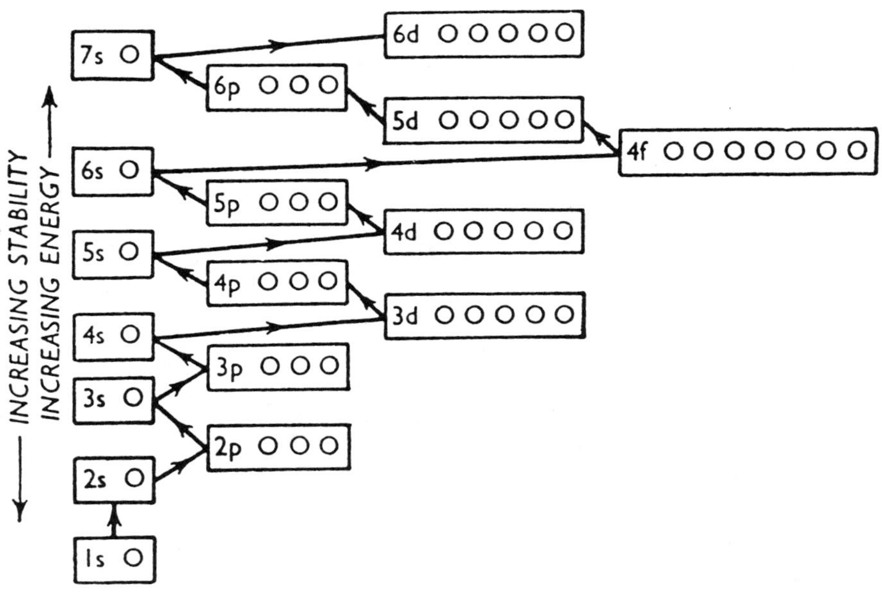

FIGURE 8. The relative energy levels of orbits in an atom.

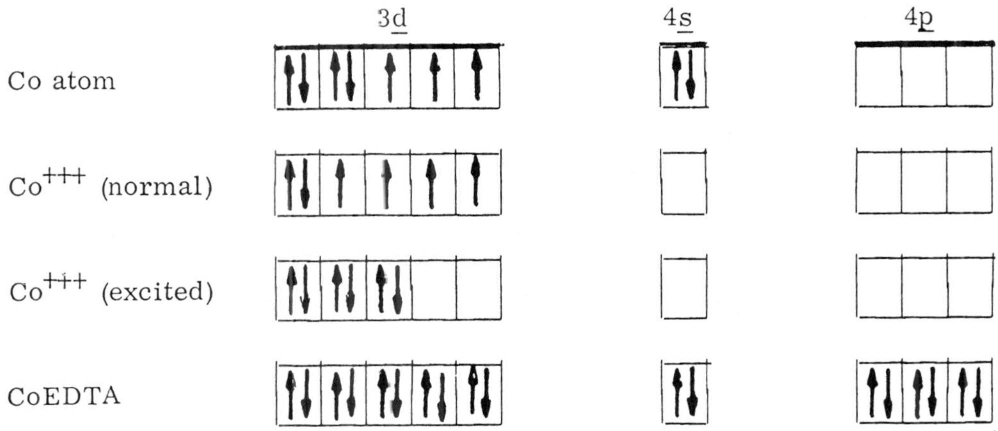

FIGURE 9. The electron distribution in Co and Co ions.

	1s	2s	2p	3s	3p	←		3d		→	4s
19 K	2	2	6	2	6						↓
20 Ca	2	2	6	2	6						↓↑
21 Sc	2	2	6	2	6	↓					↓↑
22 Ti	2	2	6	2	6	↓	↓				↓↑
23 V	2	2	6	2	6	↓	↓	↓			↓↑
24 Cr	2	2	6	2	6	↓	↓	↓	↓	↓	↓
25 Mn	2	2	6	2	6	↓	↓	↓	↓	↓	↓↑
26 Fe	2	2	6	2	6	↓↑	↓	↓	↓	↓	↓↑
27 Co	2	2	6	2	6	↓↑	↓↑	↓	↓	↓	↓↑
28 Ni	2	2	6	2	6	↓↑	↓↑	↓↑	↓	↓	↓↑
29 Cu	2	2	6	2	6	↓↑	↓↑	↓↑	↓↑	↓↑	↓
30 Zn	2	2	6	2	6	↓↑	↓↑	↓↑	↓↑	↓↑	↓↑

FIGURE 10. Electron distribution in some metals.

orbitals is increased more than of other d orbitals by ligands lying along the axis. The splitting of this energy caused by the ligand is called crystal field stablization energy (CSFE). This is illustrated in Figure 13, where the difference in energy between the two orbitals e_g and t_{2g} is for o (octahedral) crystal fields. In this pattern the algebraic sum of all energy shifts of all orbitals is zero and it is thought to preserve the center of gravity of the set of levels.[9]

IX. STABILITY CONSTANTS OR FORMATION CONSTANTS

Chelate formation constants measure the relative stability of various metal chelates. These

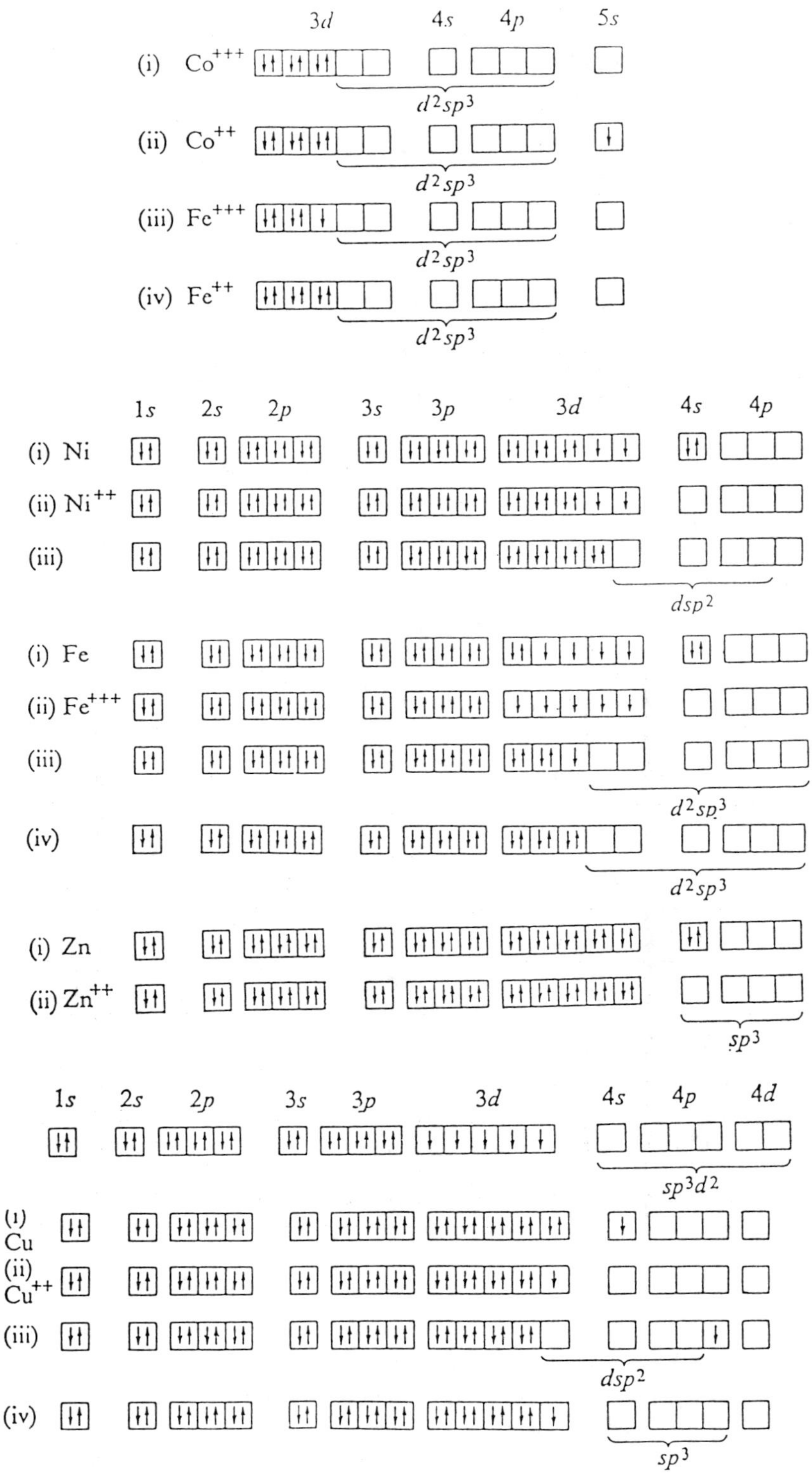

FIGURE 11. Distribution of electrons in some metals and ions.

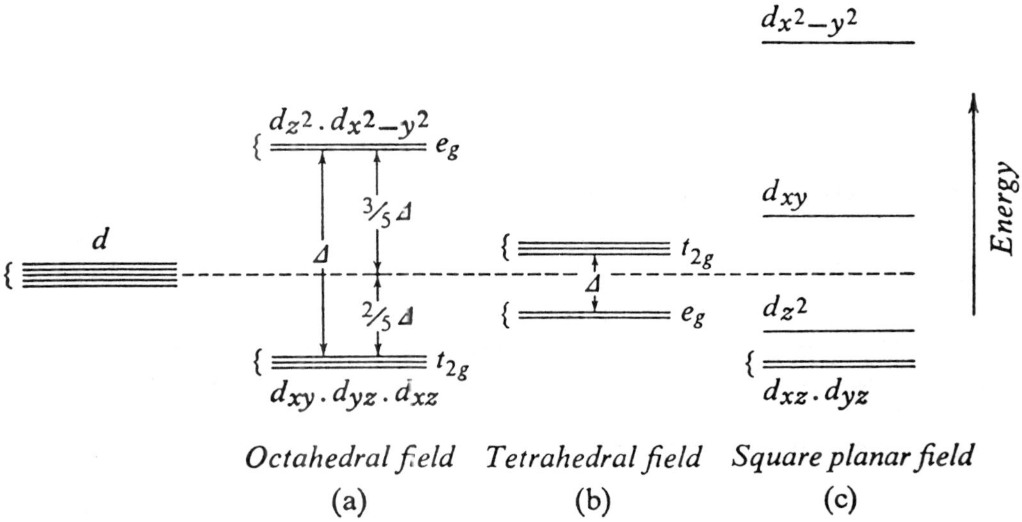

FIGURE 12. Crystal field splitting in different fields.

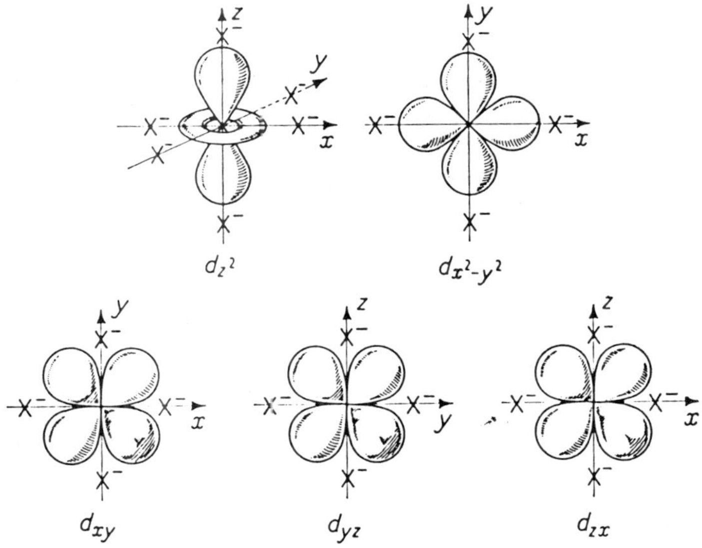

FIGURE 13. Sketches showing the electron density in the five d orbitals with respect to a set of six octahedrally arranged X^- ions.

constants indicate quantitatively the affinity of a metal to complex with a ligand. A ligand with a higher formation constant will displace a metal from a ligand with a lower formation constant for that metal. The formation of a metal chelate may be represented as displacement of hydrogen ion in a stepwise manner. The information on the theory and some of the important methods of determination of stability constants is given by Chaberek and Martell.[6]

Formation constants may be expressed in terms of concentrations (concentration constants), activities (thermodynamic constants), or in mixed (activity and concentration) terms. For a reaction $M + H \rightleftharpoons MH^+$

$$\text{Concentration constant } K_{MH} = [MH^+]/[M][H^+]$$

where [ ] = concentration.

$$\text{Thermodynamic constant } K_{MH} = [MH^+]/[M]\,[H^+] \times f_{MH^+}/f_M \cdot f_{H^+}$$

where f = fugacity or the ratio of chemical activity to concentration.

$$\text{Mixed constant } K_{MH} = [MH^+]/[M][\alpha H^+]$$

where $\alpha H^+ = [f_{H^+}] + [H^+]$ = activity = fugacity × concentration.

At zero ionic strength, an impossibility in real terms, concentration equals activity and all the formation constants become equivalent. Concentration constants may be determined and extrapolated to zero concentration.

In the absence of any precipitation of the complex, the equilibria may be attained in discreet steps as follows:

$$M + L \leftrightarrow ML \qquad K_1 = [ML]/\,[M][L]$$

$$ML + L \leftrightarrow M\,L_2 \qquad K_2 = [ML_2]/[ML][L]$$

$$ML_{n-1} + L \leftrightarrow M\,L_n \qquad K_n = [ML_n]/[ML_{n-1}][L]$$

These expressions may be written in terms of overall formation constants as follows:

$$M + L \leftrightarrow ML \qquad \beta_1 = [M\,L]/[M]\,[L]$$

$$M + 2L \leftrightarrow ML_2 \qquad \beta_1 = [M\,L_2]/[M]\,[L]^2$$

$$M = nL \leftrightarrow ML_n \qquad \beta_n = [M\,L_n]/[M]\,[L]^n$$

These two types of formation constants are related as follows where β_n is the n-th overall formation constant:

$$\beta_1 = K_1\,, \qquad \beta_2 = K_1 \cdot K_2$$

and

$$\beta_n = K_1 \cdot K_2 \cdot K_3 \cdot K_n$$

A total of n + 2 independent concentration measurements are needed to measure the value of n formation constants and the solution of such a large number of simultaneous equations is extremely tedious but practical with the help of computers. In general, the reactions may be written as follows:

$$M^{+n} + LH \leftrightarrow M\,L^{n-1} + H^+$$

$$M\,L^{n-1} + LH \leftrightarrow M\,L^{n-2} + H^+$$

$$M^{+n} + LH_m \leftrightarrow M\,L^{n-m} + m\,H^+$$

The equilibrium constant K′ for reaction is defined as:

$$K' = [M\,L^{n-m}][H^+]^m/[M^{+n}][LH_m]$$

Multiplying by $[L^{-m}]$ the numerator and the denominator:

$$K' = [M\,L^{n-m}][H^+]^m[L^{-m}]/[M^{+n}][L^{-m}][LH_m]$$

Because $M^{+n} + L^{-m} = ML^{n-m}$

Therefore,

$$K_{ML} = [M\,L^{n-m}]/[M^{+n}][L^{-m}]$$

and

$$LH_m = mH^+ + L^{-m}$$

Therefore, acid dissociation constant:

$$K_\alpha = [H^+]^m[L^{-m}]/[LH_m]$$

or

$$K' = K_{ML} \cdot K_\alpha$$

If K' and K_α could be determined experimentally, the expression would enable solution of K_{ML} or the stability constant.

The solution of a large number of simultaneous equations is required to calculate Bjerrum's stability constant. Bjerrum[0] developed a graphic method in which consecutive formation constants were determined from a plot of βn^o as a function of p[L] or the negative logarithm of the free ligand concentration. It is postulated that at $\beta_{\bar{n}} = \beta n - {}^1/_2$

$$K_{ML} = 1/[L]\,n^- - 1/2$$

The values for log K_{ML}, log K_{ML2} and log K_{ML3} are proximated from p[L] at n^- values of 0.5, 1.5, and 2.5, respectively, where $\bar{n}$ is the average number of ligand molecules bound per mole of metal.

It is not practical to measure activity coefficients of the reactants because of high ionic charges and the complicated structure of ligands, and a value of unity is assigned to activity coefficients. The concentration of the reactants, indicated by [ ], is used in calculations. Customarily, measurements for stoichiometric formation constants are conducted at a constant ionic strength by addition of an electrolyte, but could also be done over a range of concentrations and extrapolating it to zero ionic concentration when activity is equal to one.

If only one complex is formed with a known empirical formula, the formation constant can be calculated from the measurement of concentration of uncomplexed M and L, knowing the initial concentrations. The reaction is studied using polarographic, EMF, or pH measurements. A solution of a large number of simultaneous equations is required to calculate the stability constant. The numerical value of stability constants is generally given as $\log_{10}$ K. The instability constant is equal to 1/K.

In general, stability constants are measured for the model metal complexes under specified conditions when the reactants are in equilibrium and the maximum value is generally attained at extremely high, nonphysiological pH values.

X. THERMODYNAMIC CONSIDERATIONS

If ΔF^o is free energy, ΔH^o is enthalpy, and ΔS^o is entropy, then:

$$\Delta F^o = \Delta H^o - T\,\Delta S^o = -RT\ln K$$

ΔH^o can be measured for a reaction in a calorimeter.

XI. ION-EXCHANGE METHOD

An ion-exchange method was suggested by Schubert[11] for the determination of apparent formation constant K_f and not the optimal formation constant, especially at physiological pH values. In this method, Dowex-50 cation-exchanger equilibrated with 0.16 *M* NaCl solution is used. The use of this method is described by Vohra et al.[12] in their study of formation constants of Zn complexes with a number of ligands at the physiological pH of 7.4. To illustrate the basis of this method, consider a reaction between Zn ion and any complexing agent A with a valence of x as follows:

$$Zn^{2+} + n\,A^{-x} = Zn\,(A)_n^{2-xn}$$

The formation constant for this reaction is

$$K_{f(absolute)} = [Zn\,(A)_n^{2-xn}]/[Zn^{++}]\,[A^{-x}]^n$$

$$K_{f(apparent)} = K_{f(absolute)}/\alpha_H \cdot \alpha_A$$

where α_H compensates for the presence of other protonated forms of A, and α_A compensates for other complex-forming substances besides the one under investigation. Carrier-free $^{65}ZnCl$ was added to the 5,5-diethylbarbituric acid buffer of pH 7.4 to give approximately 10,000 counts/min/mℓ. The equilibration was carried out in duplicate 125-mℓ Erlenmeyer flasks. Each flask contained 200 mℓ of ion-exchange resin (sodium form), 10 mℓ of buffer solution containing ^{65}Zn, a known concentration of the chelating agent dissolved in 0.16 *M* NaCl, and the necessary amount of 0.15 *M* NaCl to make the volume 50 mℓ. The equilibration was carried out by stirring the flasks for 3 hr at about 25°C on a rotary shaker. After settling the resin, 0.5-mℓ aliquots of the supernatant solution were counted in a scintillation counter. The formation constant of the Zn complex (Y) can be then calculated as follows:

$$Y = [(K_d^O/K_d) - 1]/(A^{-x}) = K_{f1} + K_{f1} \cdot K_{f2}(A^{-x})$$

where K_d and K_d^0 are distribution coefficients for zinc in the presence and absence of the chelating agent of concentration (A^{-x}) moles per liter, respectively. K_d is calculated as follows:

$$K_d = (\%\ ^{65}Zn \text{ in exchanger})/(\%\ ^{65}Zn \text{ in solution}) \times (\text{Volume of solution (m}\ell))/(\text{Mass of ion-exchanger (mg)})$$

As the volume of solution and the mass of ion-exchanger remain constant throughout the experiment,

$$K_d = \%\ ^{65}Zn \text{ in exchanger}/\%\ ^{65}Zn \text{ in solution}$$

If Zn and the complexing agent form a complex with a 1:1 molar ratio, $Y = K_{f1}$ and is independent of (A^{-x}). If the numerical values of Y vs. (A^{-x}) are plotted on linear graph paper and a line with a finite slope is obtained, then it is assumed that 1:2 complexes also

Table 8
FORMATION CONSTANTS OF ZINC COMPLEXES OF VARIOUS LIGANDS BY ION-EXCHANGE METHOD AT A pH OF 7.4 AND CONVENTIONALLY DETERMINED VALUES

Ligand	Log formation constant: Ion-exchange[12] Kf1	Ion-exchange[12] Kf2	Conventional[6] Kf
Terephathalic acid	1.00	3.46	
Lactic acid	1.60	1.90	1.86
Thiodipropionic acid	1.91		
Sodium glycolate	1.94	2.24	1.92
Sodium orthophosphate	2.22	1.66	
Ethylenediaminebitartrate	2.60	1.60	
Citric acid	4.53		5.5
Sodium acid pyrophosphate	5.27	4.93	
Sodium hexametaphosphate	5.56	5.46	
Sodium tripolyphosphate	5.79	4.81	
Ethylenediaminediacetic acid	5.98	5.15	11.1
Hydroxyethylethylenediaminetriacetic acid	6.0	7.40	14.5
1,2-Diaminecyclohexanetetraacetic acid	6.29	6.69	18.67
Ethylenediaminetetraacetic acid	6.49	6.73	16.5
Dihydroxyethylethylenediaminediacetic acid	7.66		
Diethylenetriaminepentaacetic acid	12.27		18.14
Nitrilotriacetic acid	13.99		10.45
2-Hydroxypropylenediaminetetraacetic acid	20.08		
Ethylenediamine-*N,N'*-diacetic acid-*N,N'*-dipropionic acid	20.91		14.5

exist. The values of K_{f1} and $K_{f1} \cdot K_{f2}$ can be obtained from the intercept and the slope of the straight line, respectively. The results of determination are given in Table 8.

XII. FACTORS INFLUENCING STABILITY CONSTANTS

Stability constants are influenced by the following factors when comparisons are made under similar conditions of basicity: the nature of the donor atom, polydentate character, and resonance (Table 9).

1. Size of the ring: saturated ligands that form five membered and unsaturated ligands forming six membered rings form the most stable complexes.
2. Number of rings: as the number of rings in a chelated molecule increases, the stability of the molecule also increases.
3. Basic strength of the chelating agent: the greater the basicity of the donor ligand, the higher is its tendency to donate electrons and form stable complexes.
4. Size and charge of ligand: more stable complexes are formed by highly charged, small-sized ligands than otherwise.
5. Multidentate ligands: unless stearically hindered, form more stable complexes than monodentate ones.
6. Modification of the π-bonding strength: the central metal atom has an influence on stability and π-bonding has more influence on the stability of complexes of transition elements Ti to Cr than on complexes of Mn to Cu. If stability constants of bivalent metal complexes for transition elements are plotted against atomic numbers, the stability increases to a maximum at Cu and then decreases; the following series was obtained by Irving and Williams:[13]

Table 9
STABILITY CONSTANTS FOR SOME AMINO ACIDS

		Co^{2+}			Ni^{2+}			Cu^{2+}			Zn^{2+}			
Ligand	pK_a	log K_1	log K_2	log K_2/K_2	log K_1	log K_2	log K_1/K_2	log K_1	log K_2	log K_1/K_2	log K_1	log K_2	log K_1/K_2	log K_1 Cu-Ni
Gly^-	9.57	4.64 (−2.1)[d]	3.82	0.8	5.78 (−0.9)	4.80	1.0	8.15 (1.5)	6.88	1.3	4.96 (−1.7)	4.23	0.7	2.4
Ala^-	9.69	4.31 (−2.5)	3.50	0.8	5.40 (−1.4)	4.47	0.9	8.13 (1.4)	6.79	1.3	4.58 (−2.2)	4.00	0.6	2.7
$a\text{-}ab^-$	9.62	4.21 (−2.5)	3.50	0.7	5.35 (−1.4)	4.41	0.9	8.07 (1.3)	6.75	1.3	4.50 (−2.2)	4.15	0.4	2.7
Ser^-	9.08	4.38 (−2.0)	3.62	0.8	5.43 (−0.9)	4.53	0.9	7.89 (1.5)	6.59	1.3	4.65 (−1.7)	4.03	0.6	2.5
Thr^-	8.97	4.38 (−1.9)	3.63	0.8	5.45 (−0.8)	4.51	0.9	8.01 (1.7)	6.72	1.3	4.67 (−1.6)	3.99	0.7	2.6
Gln^-	9.00	4.04 (−2.3)	3.28	0.8	5.16 (−1.1)	4.26	0.9	7.75 (1.5)	6.48	1.3				2.6
Asn^-	8.74	4.51 (−1.6)	3.50	1.0	5.68 (−0.4)	4.55	1.1	7.86 (1.7)	6.56	1.3				2.2
Glu^{2-}	9.59	4.56 (−2.2)	3.30	1.3	5.60 (−1.1)	4.16	1.4	7.87 (1.2)	6.29	1.6	4.59 (−2.1)	3.66	0.9	2.3
Asp^{2-}	9.63	5.95 (−0.8)	4.28	1.7	7.16 (0.4)	5.24	1.9	8.57 (1.8)	6.78	1.8	5.84 (−0.9)	4.31	1.5	1.4
Phe^-	9.11	4.05 (−2.3)	3.51	0.5	5.15 (−1.2)	4.44	0.7	7.86 (1.5)	6.91	1.0	4.29 (−2.1)	4.06	0.2	2.7
Met^-	9.05	4.12 (−2.2)	3.44	0.7	5.19 (−1.2)	4.65	0.5	7.87 (1.5)	6.85	1.0	4.37 (−2.0)	3.96	0.4	2.7
SMC^-	8.73	4.12 (−2.0)	3.49	0.6	5.26 (−0.9)	4.56	0.7	7.88 (1.8)	6.84	1.0	4.46 (−1.7)	4.06	0.4	2.6
dap^o	6.66	2.91 (−2.0)			4.04 (−0.8)	3.48	0.6	6.16 (1.3)	5.12	1.0	3.20 (−1.7)	2.64	0.6	2.1
dap^-	9.39	6.28	5.08	1.2	8.16	7.00	1.2	10.51	9.32	1.2	6.30	5.35	1.0	2.4
dab^o	8.15	3.40 (−2.3)			4.54 (−1.2)	3.79	0.8	6.94 (1.2)	5.86	1.1	3.74 (−2.0)	3.34	0.4	2.4
dab^-	10.20	6.75	5.25	1.5	8.91	7.06	1.9	10.50	8.52	2.0	6.70	5.60	1.1	1.6
Orn^o	8.75	3.65 (−2.5)	3.09	0.6	4.73 (−1.4)	4.00	0.7	7.29 (1.2)	6.12	1.2	3.73 (−2.4)	3.08	0.7	2.6
Orn^-	10.52	5.01	3.48	1.5	7.11	4.92	2.2							
Lys^o	9.14	3.84 (−2.6)	3.23	0.6	4.93 (−1.5)	4.22	0.7	7.62 (1.2)	6.49	1.1	4.06 (−2.3)	3.46	0.6	2.7
Lys^-	10.67				5.75	4.59	1.2							
Arg^o	9.02	3.86 (−2.5)	3.05	0.8	4.99 (−1.3)	4.02	1.0	7.56 (1.2)	6.45	1.1				2.6
His^-	9.08	6.90 (0.5)	5.44	1.5	8.67 (2.3)	6.87	1.8	10.2 (3.8)	7.9	2.3	6.55 (0.2)	5.51	1.0	1.5

From Martin, R. B., *Metal Ions in Biological Systems*, Vol. 9, Sigel, H., Ed., Marcel Dekker, New York, 1979. With permission.

Table 10
HARD AND SOFT ACIDS AND BASES

Acids	Bases
Hard	
H^+, Li^+, Na^+, K^+	H_2O, ROH, R_2O, OH^-
Mg^{2+}, Ca^{2+}, Sr^{2+}, Mn^{2+}	OR^-, NH_3
Al^{3+}, Cr^{3+}, Fe^{3+}, Co^{3+}, Mo^{3+}	RNH_2, –NCS–, Cl–, PO_4^{3-}, SO_4^{2-}
Intermediate	
Zn^{2+}, Cu^{2+}, Ni^{2+}, Fe^{2+}	Pyridine, Br, $N_{37}NO_2^-$
Pb^{2+}, Co^{2+}	
Soft	
Cu^+, Ag^+, Au^+, Tl^+,	–SCN–, –CN–, RSH, R_2S, RS^-,
Pd^{2+}, Pt^{2+}, Pt^{4+}, Cd^{2+}	R_3P, R_3As, H^+, $S_2O_3^-$

$$Zn^{2+} < Cu^{2+} > Ni^{2+} > Co^{2+} > Fe^{2+} > Mn^{2+}$$

7. Nature of the donor ligand atom: the donor atoms may be arranged as follows:

$$H_2O > ROH > R_2O$$

$$H_3N > RNH_2 > R_2NH > R_3N$$

$$R_2S > RSH > H_2S$$

8. Resonance effect: if the resonance within the chelate ring is interfered with, the stability of the chelate may decrease. On the other hand, an increase in the resonance in the ring tends to increase the stability of the chelate.
9. Stearic effect: chelate formation can be interfered with by stearic hindrance. Less stable complexes are formed by large and bulky ligands than similar small ones. Successive stability constants tend to become smaller, suggesting a decrease in coordinating sites of the ligand due to stearic hindrance.

The metal-ligand interactions are classified in terms of hard and soft acids and bases. The examples are given in Table 10.

Metal ions may be treated as hard (small, not easily polarized), intermediate, or soft (large and easily polarized) acids (Table 11). Hard metal ions parallel the proton in their attachment to ligands, are small, and often are of high charge. They have no valence electrons that are easily distorted or removed. Soft metals ions are of low charge and their valence shell electrons are easily distorted. They bind strongly to highly polarized ligands.

Similarly, the ligands may be hard (highly electronegative), intermediate, or soft (easily polarizable) bases. Stable complexes are formed between hard acids and hard bases, or soft acids and soft bases (Table 10). In general, transition metal ions cannot be classified as hard or soft. A hard or an intermediate (borderline) acid is made softer by the coordination of a soft ligand, and harder by the coordination of a hard ligand. The metals of biological importance belong to the hard or borderline hard, and the toxic metals are in the soft group.

However, hard acids have few, not easily excited outer electrons, and soft acids have many easily excited outer electrons on donor atoms. Hard bases have high energy, inaccessible empty orbitals, and soft bases have low energy, accessible empty orbitals on donor atoms.

The preference of metal ions for ligands is indicated in Table 12.

Table 11
SOME PROPERTIES OF HARD AND SOFT ACIDS AND BASES

	Acids and bases	
	Hard	**Soft**
Polarizability	Low	High
Electropositivity	High	Low
Positive charge or oxidation state	Large	Small
Size	Small	Large
Bond types	Ionic, electrostatic	Covalent, Pi

Table 12
LIGAND GROUPS PREFERRED BY METAL IONS

Ca^{2+}	Carboxylate, phosphate, nitrogen donors
Cd^{2+}	–SH
Co^{3+}	Carboxylate, $-NH_2$, porphyrin, tyrosine
Cu^{+}	–SH as in cysteine
Cu^{2+}	Amines > carboxylate
Fe^{2+}	–SH, $-NH_2$ > carboxylate
Fe^{3+}	Carboxylate, tyrosine, $-NH_2$
Mg^{2+}	Carboxylate, phosphate, nitrogen donors
Mn^{2+}	As above
Mo^{2+}	–SH
Zn^{2+}	Imidazole, cysteine

From Hughes, M.N., *The Inorganic Chemistry of Biological Processes,* John Wiley & Sons, New York, 1982. With permission.

XIII. STUDY OF METAL COMPLEXES

The following methods are applicable to the study of metal complexes:

- Replacing one metal with another
- pH methods
- Thermodynamic measurements
- Absorption spectra
- Optical rotary dispersion and circular dichroism (CD)
- Resonance Raman spectroscopy
- Electron paramagnetic resonance (EPR)
- Nuclear magnetic resonance (NMR)
- Proton magnetic resonance (PMR)
- Mossbauer spectroscopy
- X-ray diffraction

The details of these methods will not be discussed here. Relevant information is usually developed on model compounds.

XIV. THE RELEVANCE OF STABILITY CONSTANTS IN NUTRITION

Vohra and Kratzer[15] proposed that the ability of the chelating agent to improve the

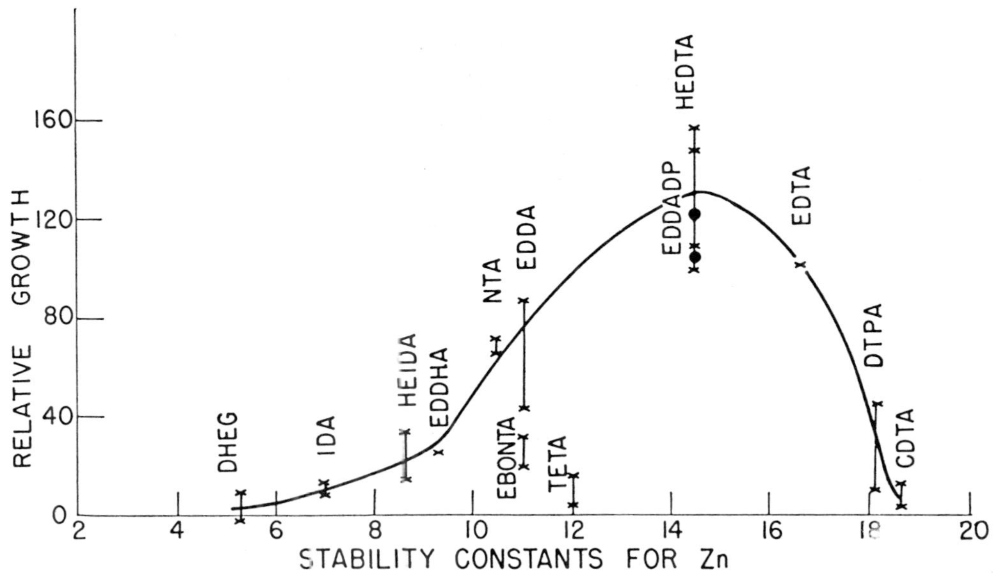

FIGURE 14. Relation of stability constant for Zn of several chelating agents to their growth promoting effect in poults when added to a Zn-deficient diet.

availability of Zn from isolated soybean protein depended upon the chelating agent having a higher stability constant for the metal than the metal-binding substance in the feed. This resulted in the loss of Zn from isolated soybean protein to the chelating agent by formation of a Zn-chelate complex in the digestive tract. The Zn-chelate was then absorbed. After absorption, the metal was available to the animal for meeting their body need if it were released from this metal-chelate complex. The implication was that ligands in the body must have a higher stability constant than the stability constant of the chelating agent releasing it from isolated soybean protein. In other words, the stability constant values of the useful chelating agent would be intermediate between the values of isolated soybean protein and the complexing agent in the body. Chelating agents were tested for their ability to improve the availability of Zn from isolated soybean protein for turkey poults. Their stability constants for Zn varied from 5.3 to 18.8. The data suggested that chelating agents with stability constant values below 13 were ineffective in removing Zn from its bound form. The stability constant of the chelating factor in soybean protein was speculated to be higher than 13, but lower than 14.5. Another chelating system in the body with a stability constant of lower than 16.5 but higher than 14 5, transported Zn to various tissues in the body where it was needed. EDDADP, HEDTA, and EDTA were active in improving the availability of Zn for poults (see Figure 14). Compounds like DTPA and CDTA would not release Zn to the endogenous chelating agent.

This was further confirmed by the studies if Nielsen et al.,[16] who tested chelating agents at a level of 1.8 mmol/kg diet on body weight, feather score, leg score, and Zn in tibias of chickens. All chelating agents with stability constants for Zn between 11.1 and 18.2, except histidine, overcame all Zn-deficiency symptoms. Histidine alleviated only bone and to a lesser extent, feather defects.

However, this hypothesis was not confirmed in a later study in which Zn was found to be equally available from Zn-complexes with citrate, EDTA, CDTA, and DTPA when fed in the diet to turkey poults (Table 13).

The availability of Zn was less from its complexes with some phosphates and phytate.[17] These studies indicate that it was not possible to predict the usefulness of a chelating agent

Table 13
INFLUENCE OF Zn FROM VARIOUS COMPLEXES ON THE GROWTH OF TURKEY POULTS OVER 21 DAYS

Treatment	Dietary Zn (mg/kg)	Weight gain (g)
Basal diet	17	36[a]*
+ ZnO	34.7	242[de]
+ Zn-phytate	40.2	219[de]
+ Zn-pyrophosphate	30.8	205[d]
+ Zn-hexametaphosphate	32.2	154[c]
+ Zn-tripolyphosphate	31.4	108[b]
+ Zn-EDTA	26.2	257[ef]
+ Zn-CDTA	31.4	258[ef]
+ Zn-DTPA	35.5	316[ef]

Note: *Asterisk indicates statistically significant differences are denoted by different letters in the column ($p <$ 0.01)

Table 14
SOLUBILITY PRODUCTS AND APPROXIMATE pHs AT WHICH 10 m*M* METAL ION SOLUTIONS FIRST PRECIPITATE

Metal ion (10 m*M*)	Solubility product (g ion $\ell^{-1})^n$	Approximate pH of onset of precipitation
Mg (II)	3.1×10^{-11}	10
Ca (II)	1.3×10^{-6}	12
Mn (II)	2×10^{-13}	9
Fe (III)	6×10^{-38}	2
Co (II)	2.5×10^{-16}	8
Cu (II)	2×10^{-19}	6
Zn (II)	1×10^{-17}	6

From Williams, D. R., *The Metals of Life,* Van Nostrand Reinhold, London, 1971. With permission.

in improving the utilization of a metal ion in animal body from the available chemical data, such as stability constants.

The stability constant values are markedly influenced by pH and the conventional values may not be very applicable in biological systems. This can be understood on the basis of data on the solubility products of metal ions in which water is treated as a ligand. At alkaline pH, complexed water produces a hydroxide ion after loss of a proton (see Table 14).

It is not possible to use theoretical stability constant values which obtained well-defined pH conditions in predicting reactions in the biological systems where the pH is different.

This is further confirmed by McCall et al.[19] in their studies with penicillamine as a chelating agent for Cu. They state that "There are no chemical data (in vitro) available at present which may be used to predict the relative effectiveness of chelating compounds in vivo. Data from the use of equilibrium stability constants (K_s), even when corrected for interfering and competing ions normally present in the body, are seldom consistent with the actual results, so evaluation of relative performance of chelating compounds is best done by in vivo tests.''[19] The relative effectiveness of EDTA and penicillamine (pen) for chelating Zn and Cu illustrates this anamoly.

Relative effectiveness (Chemical basis)	Zn-pen	>	Cu-EDTA	>	? Zn-EDTA	>>>	Cu-pen
K_S =	$10^{18.19}$		$10^{18.8}$		$10^{15.6}$		
(Actual)	Zn-EDTA	>	Cu-pen	>	Zn-pen	>>>	Cu-EDTA
K_S =	$10^{15.6}$		$10^{18.9}$		$10^{18.8}$		

One of the synthetic chelating agents, *N,N'-bis*(2-hydroxybenzyl)ethylenediamine-*N,N'*-diacetic acid (BHEBEDA), has a binding constant of 10^{40} for Fe, but is not a particularly effective Fe chelator for its removal in animal studies. However, ethylenediamine-*N,N'*-*bis*(hydroxyphenylacetic acid) (EDBHPA) and DTPA, with binding constants of 10^{34} and 10^{27}, are effective chelators for Fe in animals. The binding constant of transferrin for Fe is 10^{36}. Anderson[20] states:

> There are many reasons why a molecule can be very effective, in vitro (i.e., have a high iron binding constant) while being ineffective in vivo. First, the effective iron binding capacity under physiological conditions is very different from the iron binding capacity under optimal in vitro conditions of the metal, the chelator and other ions, particularly [H^+], [OH^-], and [Ca^{++}]. Second, the iron binding constant is an equilibrium value, not a kinetic one. Third, bioavailability and biostability are major factors. Many compounds are rapidly broken down in the bloodstream or are quickly cleared by the liver or kidneys. Oral administration adds another series of problems — the acid pH of the stomach, absorption barriers in the intestine, etc. Thus, an iron binding constant of a compound is only a rough guide for indicating potential effectiveness.

Binding constants are a good guide for the development of chelating agents for specific functions. Pitt and Martell[21] suggest that ''affinity in vivo may be defined in terms of an effective stability constant, K_{eff}, which can be estimated from the stability constant (B) of the Fe(III) complex measured under optimum in vitro conditions by correcting for the physiological pH and competing endogenous metals and ligands.'' The following equations are developed by Pitt and Martell.[21] For an equilibrium reaction between

$$Fe + L \overset{K_1}{\rightleftharpoons} FeL \qquad FeL + L \overset{K_2}{\rightleftharpoons} FeL_2 \qquad FeL_{n-1} + L \overset{K_n}{\rightleftharpoons} FeL_n$$

$$K_1 = [FeL]/[Fe][L] \qquad K_2 = [FeL_2]/[FeL][L] \qquad K_n = [FeL_n]/[FeL_{n-1}][L]$$

$$\beta_1 = K_1 \qquad \beta_n = K_1K_2 \ldots K_n$$

For effective stability constant K_{eff}, some interferences have to be considered in terms of a. Let the total uncomplexed ligand concentration be (T_L) and the fraction deprotenated be (α_L). If

$$a_L = (1 + [H^+]\beta_1^H + [H^+]^2\beta_2^H \ldots [H^+]^m\beta_m^H)^{-1}$$

If Ca and OH ions were also interfering, then

$$a_{CaL} = ([Ca]\beta_{CaL})^{-1}$$

$$a_{Fe} = (1 + [OH]\beta_1^{OH} + \ldots [OH]^m\beta_m^{OH})^{-1}$$

$$\log K_{eff} = \log B_{FeL} - n \log (a_{CaL}^{-1} + a_L^{-1}) - \log (a_{Fe}^{-1})$$

The effective solubilizing constant K_{sol} is calculated as follows:

$$\log K_{sol} = \log \beta_{FeL_n} + 1 - 3\ pH - n \log (a_L^{-1} + a_{Ca}^{-1}) + (n - 1) \log T_L$$

where the interference of other metal ions is neglected.

The information on K_{eff} and K_{sol} values is very limited. Again, these values are still measuring equilibrium constants and give no idea about the rate of establishment of equilibria. For example, the percent of Fe removal from transferrin in vitro at equilibrium was EDTA > citrate = NTA as predicted by K_{eff}. The equilibrium was established in 1 hr with citrate and NTA, and took 12 hr with EDTA. The in vivo efficiency of Fe removal is NTA > EDTA > citrate. The data on these constants should be used with caution and for guidance only.

In animal nutrition, the data developed on the basis of one ligand complexing with one metal at a time is not strictly applicable because innumerable ligands and a large number of metal ions are in a dynamic equilibrium rather than in a steady state. Information on mixed-ligand metal ion complexes might prove more useful, but still is in developmental stage. Some background to the problem can be obtained from series of publications edited by Sigel.[22]

REFERENCES

1. **Werner, A.,** *Neuere Anschauungen auf dem Gebiete der anorganischen Chemie,* 4th ed., Vieweg, Braunschweig, 1920.
2. **Morgan, G. T. and Drew, H. D. K.,** Researches on residual affinity and coordination. II. Acetylacetones of selenium and tellurium, *J. Chem Soc.,* 117, 1456, 1920.
3. **Martin, R. B.,** Complexes of α-amino acids with chelatable side chain donor atoms, in *Metal Ions in Biological Systems,* Vol. 9, Sigel, H., Ed., Marcel Dekker, New York, 1979, 1.
4. **Brown, G. I.,** *A New Guide to Modern Valency Theory,* Longman, New York, 1967.
5. **Pauling, L.,** The modern theory of valency, *J. Chem. Soc.,* 1461, 1948.
6. **Chaberek, S. and Martell, A. E.,** *Organic Sequestering Agents,* John Wiley & Sons, New York, 1959.
7. **Dweyer, F. P. and Mellor, D. P., Eds.,** *Chelating Agents and Metal Chelates,* Academic Press, New York, 1964.
8. **Van Uitert, L. G. and Fernelius, W. C.,** Coordination compounds. IX. Solution stabilities of the chelate compounds of a number of organic ligands, *J. Am. Chem. Soc.,* 76, 375, 1954.

8a. **Van Uitert, L. G. and Fernelius, W. C.,** Coordination of solution solubilities of the chelate compounds of ligands coordinating through nitrogen and oxygen, *J. Am. Chem. Soc.,* 76, 379, 1954.

9. **Cotton, F. A. and Wilkinson, G.,** *Advanced Inorganic Chemistry,* Interscience, New York, 1962.
10. **Bjerrum, J.,** *Metal Amine Formation in Aqueous Solutions,* P. Haase and Sons, Copenhagen, 1942.
11. **Schubert, J.,** Measurement of complex ion stability by the use of ion exchange resins, in *Methods of Biochemical Analysis,* Vol. 3, Glick, D., Ed., Interscience, New York, 1956, 247.
12. **Vohra, P., Krantz, E., and Kratzer, F. H.,** Formation constants of certain zinc-complexes by ion-exchange method, *Proc. Soc. Exp. Biol. Med.,* 121, 422, 1966.
13. **Irving, H. and Williams, R. J. P.,** Order of stability of metal complexes, *Nature (London),* 162, 746, 1948.
14. **Hughes, M. N.,** *The Inorganic Chemistry of Biological Processes,* John Wiley & Sons, New York, 1972.
15. **Vohra, P. and Kratzer, F. H.,** Influence of various chelating agents on the availability of zinc, *J. Nutr.,* 82, 249, 1964.
16. **Nielsen, F. H., Sunde, M. L., and Hoekstra, W. G.,** Effect of some dietary synthetic and natural chelating agents on the zinc-deficient syndrome in the chick, *J. Nutr.,* 89, 35, 1966.
17. **Vohra, P. and Kratzer, F. H.,** Influence of various phosphates and other complexing agents on the availability of zinc for turkey poults, *J. Nutr.,* 89, 106, 1966.

18. **Williams, D. R.,** *The Metals of Life,* Van Nostrand Reinhold, London, 1971.
19. **McCall, J. T., McLennan, K. G., Goldstein, N. P., and Randall, R. V.,** Copper and zinc homeostasis during chelation therapy, in *Trace Substances in Environmental Health,* Vol. 2, Hemphill, D. D., Ed., University of Missouri, Columbia, 1969, 127.
20. **Anderson, W. F.,** Iron chelation in the treatment of Cooley's anemia, in *Inorganic Chemistry in Biology and Medicine,* Martell, A. E., Ed., ACS Symp. Ser. 140, American Chemical Society, Washington, D. C., 1980, 251.
21. **Pitt, C. G. and Martell, A. E.,** The design of chelating agents for the treatment of iron overload, in *Inorganic Chemistry in Biology and Medicine,* Martell, A. E., Ed., ACS Symp. Ser. 140, American Chemical Society, Washington, D.C., 1980, 279.
22. **Sigel, H., Ed.,** *Metal Ions in Biological Systems,* Vol. 2, Marcel Dekker, New York, 1973.

Chapter 3

METAL ELEMENTS

I. INTRODUCTION

Some of the elements in the periodic table are needed in relatively large amounts by animals while others are needed in small or trace amounts. A number of the essential trace elements are transition elements. In general, the distribution of elements in the biosphere is different than in the Earth's crust, as indicated in Table 1.

Normally, plants obtain their share of elements from the soil and then make them available to the animals consuming the plants. A soil may be deficient in an element, causing a low level in the plant; this may eventually be reflected in the health of animals consuming plants grown on such soils. The element may be actually present in the plant but may not be available to the animal, thus precipitating its deficiency. On the other hand, some plants may selectively concentrate some elements, and animals consuming these plants may suffer from toxicity due to high levels of these elements.

II. BIOAVAILABILITY OF METALS

The bioavailability of mineral elements is influenced by a number of factors — both intrinsic and extrinsic. Some of these have been reviewed by Solomons[2] and O'Dell.[3] These may be classified as follows:

Intrinsic factors:
- Animal species and its genetic makeup
- Age and sex
- Monogastric or ruminant (intestinal microflora)
- Physiological function: growth, maintenance, reproduction
- Environmental stress and general health
- Food habits and nutritional status
- Endogenous ligands to complex metals (chelates)
- Extrinsic factors:
- Mineral status of the soil on which the plants are grown
- Transfer of minerals from soil to food supply
- Bioavailability of mineral elements from food to animal

The bioavailability of mineral elements from food is affected by the following considerations:
- Chemical form of the mineral (inorganic salt or chelate)
- Solubility of the mineral complex
- Adsorption on silicates, calcium phosphates, dietary fiber
- Electronic configuration of the element and competitive antagonism
- Coordination number
- Route of administration, oral or injection
- Presence of complexing agents such as chelates
- Theoretical (in vitro) and effective (in vivo) metal-binding capacity of the chelate for the element under consideration
- Relative amounts of other mineral elements

In the lumen:
- Interactions with naturally occurring ligands
- Proteins, peptides, amino acids

Carbohydrates
Lipids
Anionic molecules
Other metals
At and across the intestinal membrane:
Competition with metal-transporting ligands
Endogenously mediating ligands
Release to the target cell

The animal body contains millions of ligands. The metals are found attached to one, two or many ligands in biological systems and are of great importance for the life processes as metalloproteins, especially as enzymes. The site of biological activity often is the metal itself. The metal-ligand bonds may be as follows:[4]

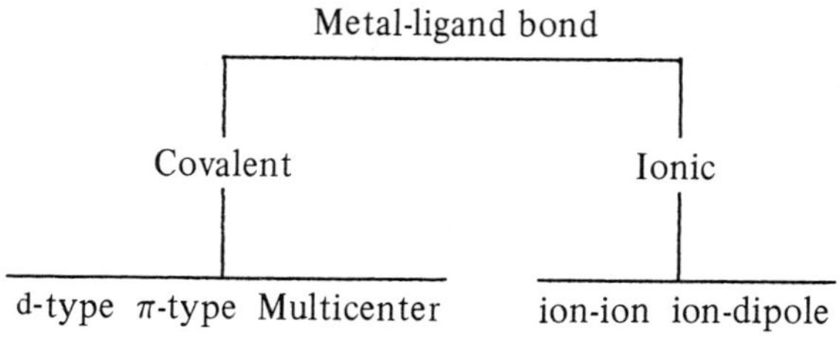

The body contains Na, K, Ca, and Mg commonly associated with skeleton and osmotic equilibrium. The concentration of K^+, Mg^{2+}, and phosphate is greater inside the cells; the concentration of Na^+, Ca^{2+}, and Cl^- is greater outside the cells. These may be grouped as two different sets. The metal ions are actively transported to maintain this gradient across the cell membrane. The enzymes activated by one set of ions may not be activated by the other set. The metal ions are bound in the chelated form selectively to ligands and are influenced by factors like ionic radii, sphere of hydration, and differences in charges. For example, Monensin, a chelating agent, selectively binds Na^+ rather than K^+.

The trace metals are part of the metalloproteins (metalloenzymes and metal-activated enzymes). The metal and the protein are combined irreversibly in the metalloenzymes. The enzyme activity is lost if the metal is removed from the protein. In the metal-activated enzymes, the metal and the protein are reversibly combined and one metal may be replaced by another metal.

Other elements (e.g., Pb, Al) enter the body as contaminants. Except for Na, K, and Ca, other elements are poorly absorbed from the GI tract.

The metal-containing enzymes contain many active sites in which a metal, generally of the transition group, is located. They are ultra-efficient catalysts capable of regenerating themselves in living systems. The activity of the enzymes depends upon three-dimensional distribution of molecules and the resultant electronic environment around these metal sites. In some of these enzymes, the metal ion is very labile and in others it is bound very strongly and its removal inactivates the enzyme. A typical representation of enzyme-metal ion-substrate bridge structure is shown in Figure 1. The schematic structures are for the ternary complexes of pyruvate kinase and manganese with fluorophosphate (A and B) and with phosphoenolpyruvate (C and D) as depicted by Mildvan et al.[5] The chelates modify this enzyme-metal ion-substrate configuration.

III. PROTEINS AS LIGANDS

The body contains millions of different proteins which are excellent ligands for metal

Table 1
DISTRIBUTION OF SOME ELEMENTS IN THE EARTH'S CRUST AND BIOSPHERE IN MILLIGRAMS PER KILOGRAM OF MATERIAL[1]

Element	Earth's crust	Sea water	Biosphere
O	4.66×10^5		a
Si	2.77×10^5	3	c
Al	8.13×10^4	1×10^{-2}	c
Fe	5.00×10^4	1×10^{-2}	c
Ca	3.63×10^4	400	b
Na	2.83×10^4	3.63×10^4	b
K	2.59×10^4	380	b
Mg	2.09×10^4	1.35×10^3	b
Ti	4.40×10^3	1×10^{-3}	
H	1.4×10^3		a
P	1.05×10^3	7×10^{-2}	a
Mn	950	+	c
F	625	1.3	c
S	260	885	b
C	200	28	a
Cl	130	1.9×10^4	b
Cr	100	+	c
Ni	75	+	c
Zn	70	+	c
As	55	+	c
Co	25	+	c
As	1.8	+	c
Mo	1.5	+	c
I	0.5	+	c
Se	1.5×10^{-2}	+	c

Note: + = present in levels below 1×10^{-2}. a = present at a level of 1 − 60%, H, C, N, O, P. b = present at a level of 0.05 − 1%, Na, Mg, S, Cl, K, Ca. c = present at a level of less than 0.05%, B, Si, Mn, Fe, Co, Cr, Ni, Cu, Zn, Se, Mo, I, Se.

ions. The proteins exhibit primary, secondary, tertiary, and quaternary structures. The polypeptide bond is the primary structure. The conformation of the polypeptide chain, such as alpha-helix and random coil, determines the secondary structure. The tertiary structure is determined by the three-dimensional configuration and is maintained by S–S linkage, hydrogen bonding, coordination to metals, and electrostatic and van der Waals forces. The subunits of a protein are its quaternary structure. The secondary, tertiary, and quaternary structures are of importance in biologically active systems. The metals form ternary complexes with proteins. These ternary complexes are active as enzymes, and are involved in the transport and storage of many other biologically active molecules.

The general role of metals in biological systems is summarized in Table 2.

A number of model systems have been studied in vitro. However, the information developed from model compounds in the laboratory cannot be applied strictly in biological systems. According to Vallee,[7] the CD and EPR spectra of model complex ions and metalloenzymes are quite different. In metalloenzymes, the metal ion is bound in low symmetry and exhibits significant distortion from the regular tetrahedral or octahedral geometries in the active state. On denaturing of the enzymes, the geometry of their coordination sites

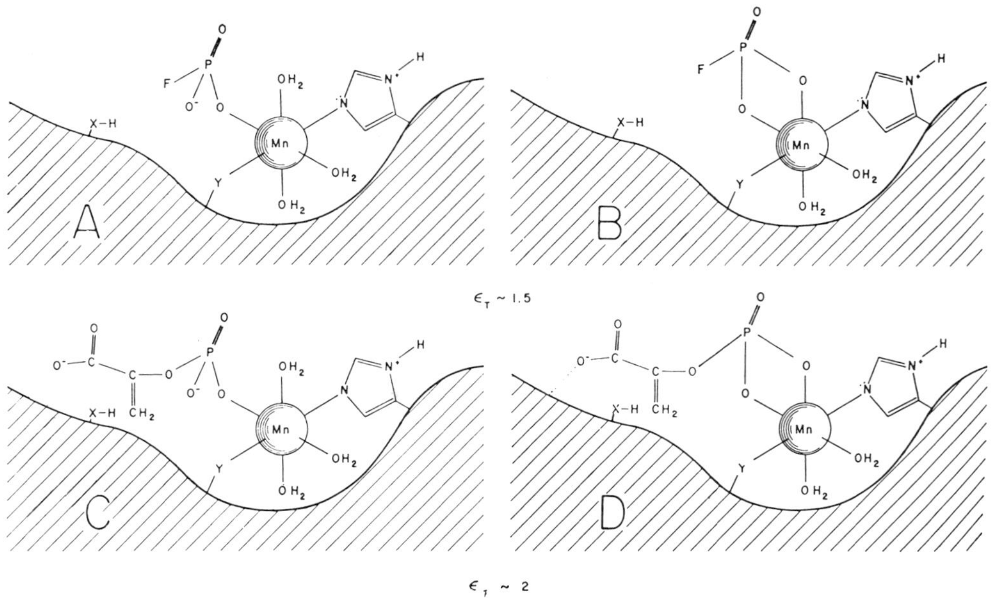

FIGURE 1. Alternative enzyme-metal ion-substrate bridge structures for the ternary complexes of pyruvate kinase and manganese with fluorophosphate (A and B) and with phosphoenolpyruvate (C and D). (From Mildvan, A. S. et al., *Biochemistry*, 6, 1805, 1967. With permission.)

Table 2
GENERAL ROLE OF METAL IONS IN BIOLOGICAL PROCESSES[6]

	Na, K	Mg, Ca, (Mn)	Zn, Cd, (Co)	Cu, Fe, Mo (Mn)
Bond strength	Weak	Moderate	Strong	Strong
Biological function	Charge transfer, nerves	Trigger reactions, hydrolysis, phosphate transfer	Hydrolysis, pH control	Oxidation-reduction reactions
Preferred ligand	O	O	N & S	N & S

becomes conventional. The binding sites of metalloenzymes are polydentate. The irregular metal-binding sites of metalloenzymes reflect an energy state which is favorable to catalysis (entatic state). Entasis (Greek) symbolizes a state of strain or stress and a reflection of the biological potential or enzyme catalysis. It differs from the strain in enzyme catalysis as it does not involve the substrate. Strain is the result of the enzyme-substate interaction. It is not easy to accurately predict the behavior of metal complexes existing in the animal body from model complexes prepared in vitro. In the body, the metal may be present as an inert metalloprotein (hemoglobin), exchangable protein-metal complex (transferrin, Ca-bound to Ca-binding protein), low molecular weight metal-ligand complex, and as aquated metal complex. The low molecular weight complexes of transition elements can pass through membranes easily and may be of zero or low-charge density.

The importance of low molecular weight ligands can be judged from the fact that Zn is present in cow's milk at a higher level (3 to 5 μg/mℓ) than in human milk (0.5 to 1.0 mg/mℓ). However, its bioavailability is much higher from human milk. Lactoferrin (~ 80,000 mol wt), a protein, is present at a concentration of about 1.5 mg/mℓ in mature human milk and only 0.02 to 0.2 mg/mℓ in cow's milk, and binds two atoms of ferric iron per mole of protein. This may have some role in improving Zn availability.[9] A number of small molecular weight compounds, such as amino acids, a polypeptide (metallothionein), *NNN'*-trimethyl-1,2-ethanolamine, prostaglandins E_2 and F_2,[10] picolinic acid, and citric acid have been proposed as Zn-binding ligands present in human milk, but citric acid[11] and picolinic acid[12] are the most implicated molecules. Other ligands, such as glutamate, may also be involved.[13] The stability constants for Zn-picolinate and Zn-citrate are 10.01 and 9.4, respectively. Zn chelated with small molecular weight chelating agents is more easily absorbed through the intestines than in combination with a large molecular complex.

For comparison, the stability constants for Cu-picolinate and Cu-citrate are 12.2 and 19.3. On the basis of stability constants of 12.2 and 10.01 for Cu-picolinate and Zn-picolinate, Sandstead[14] suggested that picolinate should also facilitate absorption of Cu by competing with ligands that form insoluble complexes with Cu in the intestinal lumen. The presence of high and low molecular weight ligands in the endogenous alimentary secretions has been suggested and the high molecular weight biliary ligands were supposed to prevent the reabsorption of Cu from bile.[15] Small ligands, such as amino acids and picolinate, could keep Cu in solution in the alkaline environment of the small intestine. However, Lönnerdal et al.[16] did not find any low molecular weight Cu complexes in human milk. Human milk was also not effective in the therapeutic treatment of Menkes' kinky hair disease.[17] On the basis of stability constants, either citric acid or picolinic acid should have bound Cu more strongly than Zn and improved Menkes' disease.

Wapnir et al.[18] studied the role of certain amino acids, dipeptides, and organic acids as ligands on the absorption of Zn by using an in vivo procedure on ileal segments of adult rats. Ligand to Zn ratios equal to or less than 3:1 were optimal for L-glutamate, glycine, L-histidine, L-tryptophan, glycylglycine, D-galacturonate, and citrate. Picolinate caused a reduction at a molar ratio of 3:1. An excess of ligand reduced Zn absorption. In this study, citrate proved better than picolinate. However, Menard and Cousins[19] found citrate and picolinate to depress Zn transport by brush border membranes vesicles rather than enhance it.

The difficulty of establishing with certainty the importance of the role of these small ligands is discussed by May et al.[20] by performing computer simulations to determine distribution of Zn among citrate, glutamate, and picolinate in human and bovine milk. At high concentrations, picolinate forms a neutral complex with Zn but at lower levels of picolinate, Zn-glutamate or Zn-citrate are the predominant low molecular weight species. Even these data are not in agreement with the data of Holt,[21] suggesting negligible formation of the Zn-picolinate and the protinated Zn-citrate species.

The suggested involvement of prostaglandins in improving Zn absorption[22] was not confirmed by studies of Meydani et al.[23] in rats.

The naturally occurring chelating agents differ from synthetic agents like EDTA. For example, a histidine-Zn complex is highly absorbed as compared to the EDTA-Zn complex and its absorption is suppressed by 2,4-dinitrophenol but not of EDTA-Zn complex. The EDTA-Zn complex is a charged molecule in contrast to the neutral histidine-Zn complex. In general, the ligand-metal reaction takes place at a relatively constant temperature and pH in the body, and the pH changes in in vitro systems.

IV. GENERAL THEORY OF METAL ABSORPTION

The metal is ingested in the food as a salt or as part of plant and animal tissues. It may be complexed in the tissues with ligands, either organic or inorganic or a combination of both. More than one metal may be at different sites in these ligands. It may be in the form of metalloproteins and low molecular weight and high molecular weight chelates. In the GI tract, the endogenous secretions react on the food during its digestion. Also, some metals are part of these endogenous secretions. The lumen contains free metal, low and high

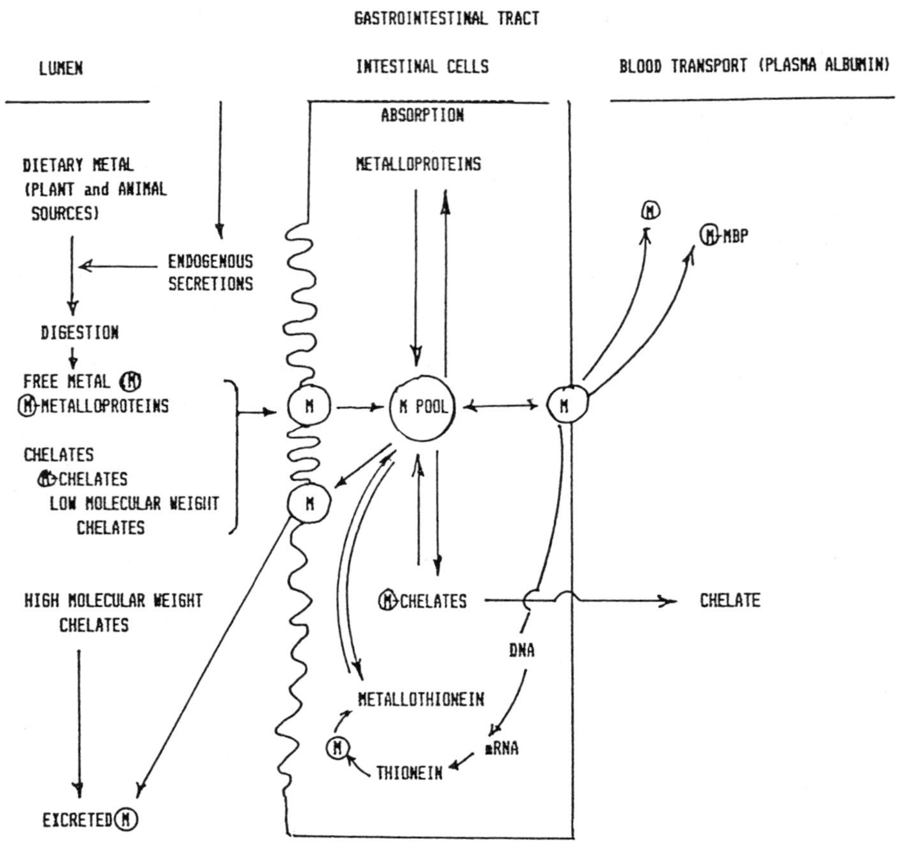

FIGURE 2. Absorption of metals through the intestinal cells involving chelates including metalloproteins, and metal-binding proteins (MBP).

molecular weight chelates of the metal, metalloproteins, and various ligands. Depending upon the effective stability constants of the ligands and the metal, the concentration of other metals, the insolubility, and adsorption considerations, new dynamic equilibria are established to form new metal complexes. The low molecular weight metal complexes cross the intestinal cell wall. The metal-chelate is in equilibrium with free metal pool, metalloproteins, metallothionein, and other chelates. The synthesis of metallothionein is dependent upon DNA-mediated messenger RNA and is controlled by the concentration of the metal circulating in the blood. The metal is transported from or to the intestinal cell in the form of complexes by metal-binding proteins. The movement of metal is thus regulated from the lumen to blood plasma albumin as well as in the opposite direction. The chelate is eventually metabolized or excreted. This is depicted in Figure 2.

V. FACTORS INFLUENCING METAL BIOAVAILABILITY

The topic of bioavailability of metals has been reviewed by O'Dell,[24] Shah,[25] and Solomons,[2] especially for humans. Actually, different species of animals may not have the same bioavailability of a metal. The data from rats may not be applicable to chickens, guinea pigs, or other species and vice versa. The dietary components are also important. The data obtained with diets based on isolated soybean protein may differ from the data with diets containing casein or egg albumin.

Table 3
EFFECT OF SOME LIGANDS ON CALCIUM BIOAVAILABILITY

Decrease	Increase	No effect
Phytate	Lactose	Phosphorus
Fiber	Medium chain triglycerides	Protein
Oxalate	Vitamin D	Ascorbic acid
Fat		Citric acid
Alcohol		Pectin
Sodium a ginate		
Uronic acid		
EDTA?		
NTA?		

The role of unavailable carbohydrates as chelating agents is also important. For example, Mg and Zn absorption was significantly reduced in animals consuming 10 and 20% cellulose compared with animals ingesting 0 or 5% cellulose in their diets.[26]

In the testing of the effect of a chelating agent on the performance of an animal model, only three types of results are possible. The performance may be improved, it may be depressed, or it may not be affected at all. The proper parameters for the measurement of performance need development. Survival and growth are important, but other biochemical parameters are needed for further confirmation of the actual role of the metal under study and the mechanism involved. These are hard to pinpoint.

A. Ca

The availability of Ca is influenced by a number of factors.[27] The solubility of Ca-amino acid complexes was not very important in the absorption of Ca. Lysine and arginine, but not glycine, promoted absorption of Ca and Mg in rats in spite of the high solubility of the Ca-glycinate complex. Information on some factors influencing Ca availability is given in Table 3.

An excess of vitamin D and a deficiency of P also increase Ca absorption.

B. Zn

Solubility and complex formation play a role in Zn availability. Zn is equally available from sulfate, oxide, and carbonate. Diets based on plant seeds and cereals have produced Zn deficiency in pigs and poultry in the presence of relatively high levels of Ca. EDTA improves the availability of Zn under these conditions. The level of amino acids influences Zn availability. High levels of histidine reduce and low levels increase absorption through ligated intestinal segments of rats. The effect of some factors on Zn bioavailability is summarized in Table 4.

VI. METAL SUPPLEMENTS

The essential nature of metals for a living system is established beyond doubt. The metal concentration is controlled within narrow limits in the body for optimal well-being by proteins and hormones. If the controls fail, disorders or diseases are likely to arise. The disorders may be due to a deficiency or an excess of the metal, as well as interactions between cations and anions. The excesses cause toxicity. The ratio between requirement and toxicity may be narrow or wide. This is illustrated in Figure 3. The metals may be classified as follows:

1. Essential metals not held firmly in coordination compounds: Na, K, Mg, Ca

Table 4
FACTORS INFLUENCING ZINC BIOAVAILABILITY

Increase	Decrease	No effect
EDTA	Phytate	Ascorbic acid
Citric acid?	Phosphates	
Picolinic acid?	Protein complexes	
Amino acids?	Amino acids?	
	Microflora	
	High Ca	

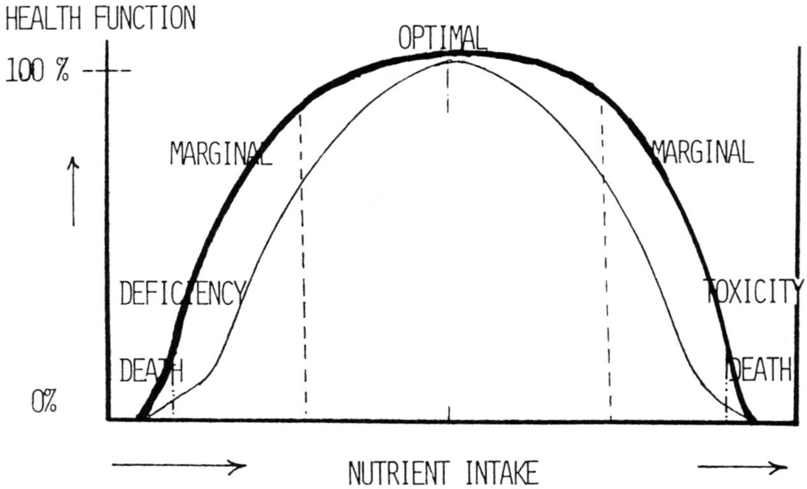

FIGURE 3. Two extreme dose-response curves.

2. Essential metals present as complexes: V, Cr, Mn, Fe, Co, Ni, Cu, Zn, Mo
3. Present in plants and animal tissues and may have a biological role: Rb, B, Al, Sr, Ba
4. Metals (contaminant) without any known function: Ag, Pb
5. Not detected in newborns but accumulate during life as contaminants: Cd, Sn, Pb, Au, Bi, Ga, Hg, Pt

The metal ions are generally complexed with ligands in the body. The nature and size of these ligands is highly variable. More information is available on high molecular weight metalloproteins occurring in relatively large amount than on low concentration, low molecular weight ligands in a state of dynamic equilibrium.

The criteria for relative availability of the metal from the diet must be well defined. The usual criteria are bioassays with animals considering survival, growth, overcoming of deficiency symptoms, function of specific metalloenzymes, and tissue profile for the metals. The metal-complexes serve many diverse functions, such as:

1. Transfer of materials such as oxygen by hemoglobin
2. Energy transfer by chlorophyll in plants and by cytochromes
3. Metalloenzymes
4. Metal storage — metalloproteins such as ferritin (445,000 mol wt) hold 4500 atoms of Fe^{3+}; hemosiderin, 37% Fe; ceruloplasmin (160,000 mol wt) holds 8 atoms of Cu

The metals may be already present in diets as an integral part of the ingredients or may be supplemented as inorganic salts or in combination with ligands. The choice depends upon many factors, such as the amount and type of cation needed, the anions involved, the presence of other cations and their levels, the availability of cations to the animal, the solubility of the supplement in the environment of the digestive tract, and the cost of the supplement. The pH in the intestinal tract varies from about 2 to about 8, but the blood pH is 7.4.

The metals may be added as oxide, carbonate, sulfate, phosphate, and chlorides. However, zinc chloride and ferric chloride are moderately irritant. These salts readily release metal ions. If added to food, the metal ions from these supplements promote oxidation of fat-soluble vitamins and unsaturated fatty acids. This is an undesirable property. The ions are not readily released from metal-ligand complexes without a drastic change in pH. The addition of metal-ligand complexes is more desirable for a long shelf-life of the food. The most commonly used ligands are citric acid, picolinic acid, lactic acid, tartaric acid, amino acids, peptides, fructose, EDTA, NTA, and other synthetic chelates. In general, a chelating agent must be of a relatively low molecular weight. This rules out protein complexes. A good target is a molecular weight of not more than 1000. The chelate should have an effective stability constant higher than that of the food ligand to remove it from food, cross the intestinal cell with or without the metal, and have a lower stability constant than of those ligands in the transport system which delivers the metal to the target tissue. The ligand should be eliminated from the body by excretion as such or of its metabolic products. The ligand should not precipitate a deficiency of some other essential metal, but may complex and excrete any toxic levels of other metals.

A metal element may be absorbed through the intestinal tract in bound form, but then excreted without being available for utilization by the animal. Fe in the form of the EDTA complex is readily absorbed through the intestinal tract, but is excreted in the urine, and is not utilized for prevention of anemia.

The relative availability of a metal element from any source is not the same for all species of animals. For example, Co is needed for the synthesis of vitamin B_{12}, an essential nutrient for all animals. It is incorporated by the microorganism present in the rumen of ruminants into vitamin B_{12}. Unless they practice coprophagy, monogastric animals need vitamin B_{12} in their diet rather than Co. Co salts or chelates other than vitamin B_{12} are sufficient for ruminants, but useless for humans, poultry, and pigs.

Fe may be supplied to monogastric animals in ferrous or ferric form as oxide, carbonate, sulfate, phosphate, phytate, chloride, fumarate, gluconate, citrate, ammonium citrate, complexed with fructose, NTA, EDTA, or as heme. Fe is absorbed from all the additives which are easily soluble. However, heme Fe (616.5 mol wt) is more available than nonheme Fe. Ascorbic acid promotes absorption of nonheme Fe and not heme Fe. Some low molecular weight chelating agents, like tannins and phytic acid, as well as large molecular weight egg white and milk proteins, reduce Fe availability. More Fe is absorbed from NaFe·EDTA than from $FeSO_4$, but the availability of Fe from NaFe·EDTA to target tissues has to be confirmed.

In a study with humans, citric acid reduced the absorption of nonheme Fe from diets consisting of maize, rice, and black beans, but protein from meat or soybeans, and ascorbic acid improved it. The choice of a supplement depends upon many factors, including cost.[28,29] Like Fe, Zn and Cu are also readily available from their soluble salts and chelates.

If the price is favorable, it is desirable to supplement metals in the form of low molecular weight chelates because they are less likely to promote oxidative rancidity in foods and spoil fat-soluble vitamins and unsaturated fatty acids. Chelation of metals imparts greater solubility. The ligands which can be easily metabolized in the body are more desirable than the poorly metabolizable ones. The latter are more useful in the elimination of excessive levels of metals from the body and prevent toxicity.

The question arises about the preferred complexes for absorption and excretion. A metal may be transported across the intestinal membrane by active transport or by passive diffusion. The active transport may be involved in metal-amino acid complexes. The rates for passive diffusion are dependent upon equilibrium considerations. The metal may be present in the digestive tract in the following forms:[30]

1. Inert metalloprotein such as Fe in hemoglobin which would not release Fe without destruction of the complex
2. Metal exchangeable from a protein complex as transferrin by another ligand
3. Low molecular weight complexes which are easily diffusible
4. A low concentration of metal ions in equilibrium with other complexes

The easily diffusible low molecular weight ligands of low or zero charge are desirable. Also, a concentration gradient must exist on the two sides of the membrane. Microparticles of solids may also diffuse between the cells while large insoluble complexes cannot diffuse. A low molecular weight complex may have a large sphere of hydration which may hinder its diffusion.

Passive diffusion is facilitated by lipophylic ligands. A fully ionized metal-ligand complex in the intestinal tract would need a specific carrier to cross the lipoprotein epithelial membrane but a nonionized metal-ligand complex will pass easily. Pitt and Martell[31] suggest that the preferred ligands would be:

1. Lipophilic
2. The pK_a of any acidic group(s) in the molecule should be >3, and of basic groups <8
3. The availability of specific carrier

A long alkyl chain imparts lipophilic properties. The pK is more difficult to manipulate because it may reduce the function of the ligand.

The low molecular weight, neutral or low charge density metal complexes may be absorbed by the target tissue or excreted through the bile duct. Charged complexes are excreted through the kidney preferably. For example, 2,3-dimercapto-1-propanol (BAL) promotes Cd excretion through the bile duct, and sodium dimercaptopropanesulfonate (DMPS) through kidney because of the charge difference.

A few ligands of choice are as follows:

Fructose, glucose, sucrose
Lactic acid, citric acid, tartaric acid, gluconic acid, ascorbic acid, glycolic acid, hydroxybutyric acids, glyceric acid, malic acid
Amino acids especially glycine, glutamic acid, histidine, methionine, lysine
Peptides
Phosphates
Porphyrins
Some synthetic chelates
DTPA, EDDHA, *bis*-(hydroxyethyl)-glycine, EDTA, HEDTA, NTA

VII. USE OF ORBITAL THEORY IN NUTRITION

Hill and Matrone[32] fed a skim milk-glucose diet in their studies with chickens to study trace element interactions and observed antagonism between certain trace elements. Their results are given in Tables 5 to 7.

Table 5
INTERACTION OF Zn AND Cu IN CHICKS

	Cu (mg/kg)			
Zn (mg/kg)	0	10	0	10
	Hemoglobin, g/100 mℓ		Body wt (g)	
0	6.5	8.5	262	272
100	6.0	8.8	182	294
200	4.8	8.0	139	291
300	4.1	6.6	127	310

Table 6
INTERACTION OF Cd WITH Zn AND Cu

		Cu (mg/kg)		
Cd (mg/kg)	Zn (mg/kg)	0	10	Significant effects
		Body wt (g)		
0	0	214	246	Cu × Cd
100	0	131	163	Zn × Cd
0	200	174	222	
100	200	149	197	

Table 7
INTERACTION OF Cu WITH Hg AND Zn

		Cu (mg/kg)	
		0	25
		Body wt (g)	
Hg (mg/kg)	0	100	130
	400	104	111
Zn (mg/kg)	0	107	119
	400	95	123

Hill and Matrone[32] and Matrone[33] have used the theory based on orbitals in predicting trace-element antagonism between Cu and Zn, Fe and Cu, and Cd and Zn, and between sulfate and molybdate. The chemical parameters for these cations and anions are given in Table 8.

According to Matrone,[33] "copper would likely be present in both ionic forms, Cu^{+} (d^{10}) and Cu^{2+} (d^{9}). Accordingly, metal ions with either a tetrahedral (sp^3) or a square planar (dsp^2) complex configuration might be expected to be antagonistic to copper in vivo." As Zn has the same chemical parameters as Cu, it interfered with metabolism of Cu. Cd may be toxic because it has similar chemical parameters as Zn and competes with it. For the same reasons, arsenate and vandate, but not chromate and selenate, inhibited phosphate uptake by respiring mitochondria. Arsenate and vandate have an sp^3 configuration and one nd like that of phosphate anion; however, chromate and selenate have an sp^3 but two nd bonds.

Table 8
CHEMICAL PARAMETERS FOR SOME CATIONS AND ANIONS[33]

Cation	Orbital	Configuration	Coordination no.
Cu^{+}	d^{10}	sp^3	4
Zn^{2+}	d^{10}	sp^3	4
Cd^{2+}	d^{10}	sp^3	4
Hg^{2+}	d^{10}	sp	2
Cu^{2+}	d^9	dsp^2	4
Ag^{2+}	d^9	dsp^2	4
Anions	**No. of π d binds**		**Configuration**
PO_4^{3-}	1		sp^3
VO_4^{3-}	1		sp^3
AsO_4^{3-}	1		sp^3
MoO_4^{2-}	2		sp^3
CrO_4^{2-}	2		sp^3
SeO_4^{2-}	2		sp^3
SO_4^{2-}	2		sp^3

VIII. METABOLISM OF CHELATES

Most of the chelating agents normally present in the body are in a dynamic state undergoing metabolism. However, the information on the metabolism of synthetic chelating agents is extremely limited. Also, the rate of their absorption and metabolism may not be uniform in all the species. A maximum of 18% of an orally administered dose of Ca-EDTA-^{14}C was absorbed by rats in 24 hr and less than 0.1% was respired by the rats. After a parenteral administration, 95 to 98% of the EDTA administered was excreted in the urine in 6 hr.[34] In humans, the absorption was about 5%.[35] Stevens et al.[36] observed that following an oral administration to humans, 90 to 100% of DTPA-^{14}C or EDTA-^{14}C passed out unabsorbed, but the presence of some ^{14}C in the urine indicated some absorption of the chelating agents. Intravenously injected ^{14}C-labeled chelates were excreted quantitatively in the urine in 24 hr.

Darwish and Kratzer[37] administered a solution containing 0.0063 mmol of $^{65}ZnCl_2$ and 0.0074 mmol EDTA-2-^{14}C in gelatin capsules to two colostomized hens fed a Zn-deficient diet. The original ^{14}C to ^{65}Zn ratio was 2:2. Their data are given in Table 9.

These data suggest that EDTA was absorbed in the hens at a faster rate than Zn in the initial period, but subsequently Zn accumulated and EDTA was excreted or metabolized as indicated by the ratios. EDTA was confined to plasma and did not penetrate red blood cells. Zn was present in cells and plasma. EDTA was metabolized and the ^{14}C activity in respired gas peaked at 7 and 30 hr, but some activity was present even at 110 hr, possibly as a result of utilization of secondary metabolites. This led to the hypothesis that EDTA was removing Zn from its complexes with isolated soybean protein and forming a Zn-EDTA complex. The size of Zn-EDTA complex was possibly small enough to be absorbed through the GI tract. In blood and tissues, EDTA released Zn to proteins and enzymes with a greater effective binding constant for Zn than EDTA.

In the studies of Koike et al.[38] on the absorption of ^{65}Zn-EDTA-^{14}C or ^{45}Ca-EDTA-^{12}C complexes placed in the lumen of a chicken, ^{14}C and ^{65}Zn or ^{45}Ca appeared in the blood, perfusing the isolated segment of the small intestines. The ^{14}C to ^{65}Zn ratios in the luminal contents divided by similar ratios of the injected complexes gave values >1.0, indicating a

Table 9
^{14}C TO ^{65}Zn RATIO IN BLOOD AFTER AN ORAL ADMINISTRATION TO COLOSTOMIZED HENS

Time after administration	Ratio
Initial	2.2
Hour:minute	
0:45	5.0
1:15	6.0
4:45	0.5
8:00	0.65
12:00	1.3
24:00	1.3
36:00	1.0
48:00	1.0

greater absorption of EDTA than of Zn and suggesting an adsorption of Zn by the lumen walls. In contrast to this, ^{14}C from EDTA-^{45}Ca complex was transferred even more rapidly through the mucosa than ^{45}Ca, as shown by a lower ^{14}C to ^{45}Ca ratio in the blood. The recovery of Ca was two to three times greater than that for Zn over comparable periods, suggesting no luminal adsorption of Ca.

Sahagian et al.[39] measured the transmural movements of Zn, Mn, Cd, and Hg by isolated segments of rat's small intestine. The relative rate of increase in the concentration of these metals in the serosal solution, or a measure of rates of transmural passage was Mn > Cd > Hg > Zn, and the tissue metal intake was Zn > Hg > Cd > Mn. In the presence of EDTA, the transmural movement of Mn, Zn, and Cd was greatly enhanced but was depressed in the case of Hg, while ascorbic acid had no effect on Mn, and enhanced that of Cd and Zn.

REFERENCES

1. **Ochiai, E.,** *Bioinorganic Chemistry, An Introduction,* Allyn and Bacon, Boston, 1977.
2. **Solomons, N. W.,** Biological availability of zinc in humans, *Am. J. Clin. Nutr.,* 35, 1048, 1982.
3. **O'Dell, B. L.,** Bioavailability of essential and toxic trace elements, *Fed. Proc. Fed. Am. Soc. Exp. Biol.,* 42, 1714, 1983.
4. **Craig, D. P. and Nyholm, R. S.,** The nature of the metal-ligand bond, in *Chelating Agents and Metal Chelates,* Dweyer, F. P. and Mellor, D. P., Eds., Academic Press, New York, 1964, 51.
5. **Mildvan, A. S., Leigh, J. S., and Cohn, M.,** Kinetic and magnetic resonance studies of pyruvate kinase. III. The enzyme-metal-phosphoryl bridge complex in the fluorokinase reaction, *Biochemistry,* 6, 1805, 1967.
6. **Williams, D. R.,** *The Metals of Life,* Van Nostrand Reinhold, London, 1971.
7. **Vallee, B. L.,** The entatic properties of cobalt carboxypeptidase and cobalt procarboxypeptidase, in *Trace Element Metabolism in Animals,* Vol. 2, Hoekstra, W. G., Suttie, J. W., Ganther, H. E., and Mertz, W., Eds., University Park Press, Baltimore, 1974, 5.
8. **Williams, D. R.,** Trends towards the study of low molecular weight complexes in biological systems, in *Coordination Chemistry,* Banerjea, D., Ed., Pergamon Press, New York, 1980, 129.
9. **Rumball, S. V., Ainscough, E. W., Brodie, A. M., Plowman, J. E., Baker, H. M., and Brooks, H. J. L.,** Zinc-binding ligands in human milk, in *Trace Element Metabolism in Man and Animals (TEMA-4),* Howell, J. M., Gawthorne, J. M., and White, C. L., Eds., Springer-Verlag, Basel, 1982, 253.

10. **Cousins, R. J.,** Regulation of zinc absorption: role of intracellular ligands, *Am. J. Clin. Nutr.*, 32, 339, 1979.
11. **Lönnerdal, B. L., Stanislowski, A. G., and Hurley, L. S.,** Isolation of low molecular weight zinc binding ligand from human milk, *J. Inorg. Biochem.*, 12, 71, 1980.
12. **Evans, G. W. and Johnson, P. E.,** Characterization and quantitation of a zinc-binding ligand in human milk, *Pediatr. Res.*, 14, 976, 1980.
13. **Sandstead, H. H.,** Copper availability and requirements, *Am. J. Clin. Nutr.*, 35, 808, 1982.
14. **Gollan, J. L.,** Studies on the nature of complexes formed by copper with human alimentary secretions and their influence on copper absorption in the rat, *Clin. Sci. Mol. Med.*, 49, 237, 1975.
15. **Martin, M. T., Licklider, K. F., Brushmiller, J. G., and Jacobs, F. A.,** Detection of low molecular weight copper (II) and zinc (II) bindings ligands in ultrafiltered milks — the citrate connection, *J. Inorg. Biochem.*, 15, 55, 1981.
16. **Lönnerdal, B., Keen, C. L., Hoffman, B., and Hurley, L. S.,** Copper ligands in human milk — a vehicle for copper supplementation in the treatment of Menkes' disease?, *Am. J. Dis. Child.*, 134, 802, 1980.
17. **Williams, D. M., Atkin, C. L., Seay, A. R., and Bray, P. F.,** Failure of human milk therapy in Menke's kinky hair disease, *Am. J. Dis. Child.*, 133, 218, 1979.
18. **Wapnir, R. A., Khani, D. E., Bayne, M. A., and Lifshitz, F.,** Absorption of zinc by the rat ileum: effects of histidine and other low-molecular-weight ligands, *J. Nutr.*, 113, 1346, 1983.
19. **Menard, M. P. and Cousins, R. J.,** Effect of citrate, glutathione and picolinate on zinc transport by brush border membrane vesicles from rat intestine, *J. Nutr.*, 113, 1653, 1983.
20. **May, P. M., Smith, G. L., and Williams, D. R.,** Computer calculation of zinc(II)-complex distribution in milk, *J. Nutr.*, 112, 1990, 1982.
21. **Holt, C.,** Zinc binding ligands in milk: both arguments err seriously (letter to the editor), *J. Nutr.*, 111, 2240, 1981.
22. **Evans, G. W. and Johnson, P. E.,** Defective prostaglandin synthesis in acrodermatitis enteropathica, *Lancet*, 1, 52, 1977.
23. **Meydani, S. N., Meydani, M., and Dupont, J.,** Effects of prostaglandin modifiers and zinc deficiency on possibly related functions in rats, *J. Nutr.*, 113, 494, 1983.
24. **O'Dell, B. L.,** Dietary factors that affect biological availability of trace elements, *Ann. N.Y. Acad. Sci.*, 199, 70, 1972.
25. **Shah, B. G.,** Chelating agents and bioavailability of minerals, *Nutr. Res.*, 1, 617, 1981.
26. **Gordon, D. E., Besch-Williford, C., and Ellersieck, M. R.,** The action of cellulose on the intestinal mucosa and element absorption by the rat, *J. Nutr.*, 113, 2545, 1983.
27. **Allen, L. H.,** Calcium bioavailability and absorption: a review, *Am. J. Clin. Nutr.*, 35, 783, 1982.
28. **Hallberg, L.,** Bioavailability of dietary iron in man, *Ann. Rev. Nutr.*, 1, 123, 1981.
29. **Hallberg, L. and Rossander, L.,** Improvement of iron nutrition in developing countries: comparison of adding meat, soy protein, ascorbic acid, citric acid and ferrous sulfate on iron absorption from a simple Latin American-type of meal, *Am. J. Clin. Nutr.*, 39, 577, 1984.
30. **May, P. M. and Williams, D. R.,** Complex speciation for absorption and excretion, in *Trace Element Metabolism in Man and Animals (TEMA-4)*, Gawthorne, J. M., Howell, J. McC., and White, C. L., Eds., Springer-Verlag, Basel, 1982, 149.
31. **Pitt, C. G. and Martell, A. E.,** The design of chelating agents for the treatment of iron overload, in *Inorganic Chemistry in Biology and Medicine*, Martell, A. E., Ed., ACS Symp. Ser. 140, American Chemical Society, Washington, D.C., 1980, 279.
32. **Hill, C. H. and Matrone, G.,** Chemical parameters in the study of in vivo and in vitro interactions of transition elements, *Fed. Proc. Fed. Am. Soc. Exp. Biol.*, 29, 1474, 1970.
33. **Matrone, G.,** Chemical parameters in trace-element antagonism, in *Trace Element Metabolism in Man and Animals*, Vol. 2, Hoekstra, W. G., Suttie, J. W., Ganther, H. E., and Mertz, W., Eds., University Park Press, Baltimore, 1974, 91.
34. **Foreman, H., Vier, M., and Magee, M.,** The metabolism of C^{14}-labeled ethylenediaminetetraacetic acid in the rat, *J. Biol. Chem.*, 203, 1045, 1953.
35. **Foreman, H.,** The pharmacology of some useful chelating agents, in *Metal Binding in Medicine*, Seven, M. G. and Johnson, L. A., Eds., Lippincott, New York, 1960.
36. **Stevens, E., Rosoff, B., Weiner, M., and Spencer, H.,** Metabolism of the chelating agent diethylenetriamine pentaacetic acid (C^{14}DTPA) in man, *Proc. Soc. Biol. Med.*, 111, 235, 1962.
37. **Darwish, N. M. and Kratzer, F. H.,** Metabolism of ethylenediaminetetraacetic acid (EDTA) by chickens, *J. Nutr.*, 86, 187, 1965.
38. **Koike, T. I., Kratzer, F. H., and Vohra, P.,** Intestinal absorption of zinc or calcium-ethylenediaminetetraacetic acid complexes in chickens, *Proc. Soc. Exp. Biol. Med.*, 117, 483, 1964.
39. **Sahagian, B. M., Harding-Barlow, I., and Perry, H. M., Jr.,** Transmural movement of zinc, manganese, cadmium and mercury by rat small intestine, *J. Nutr.*, 93, 291, 1967.

Chapter 4

CONSIDERATION OF SPECIFIC CHELATING AGENTS: ROLE OF PHYTIC ACID AND OTHER PHOSPHATES AS CHELATING AGENTS

I. GENERAL CONSIDERATION

Orthophosphoric acid has three functional OH groups attached to P and is a good chelating agent. The phosphate molecules may be condensed in anhydride linkage after loss of water to form linear or cyclic polymers. The examples of these polymers are pyrophosphoric acid, tripolyphosphoric acid, trimetaphosphoric acid, and tetrametaphosphoric acid. Some of these acids are extremely unstable and can only be prepared as salts by heating Na or K under controlled conditions. These salts are chelating agents and are used as water softeners. The structures of the some of these phosphates are as follows:

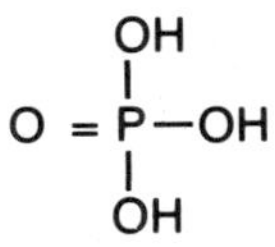

Phosphoric acid

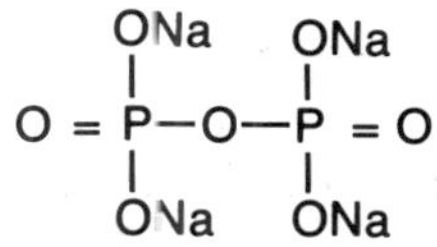

Sodium pyrophosphate

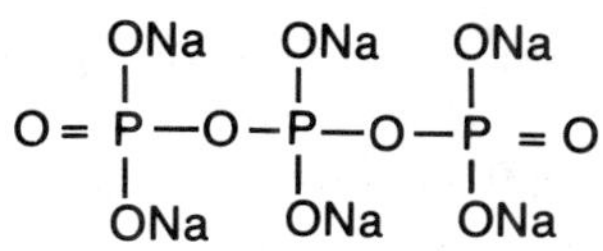

Sodium tripolyphosphate acid

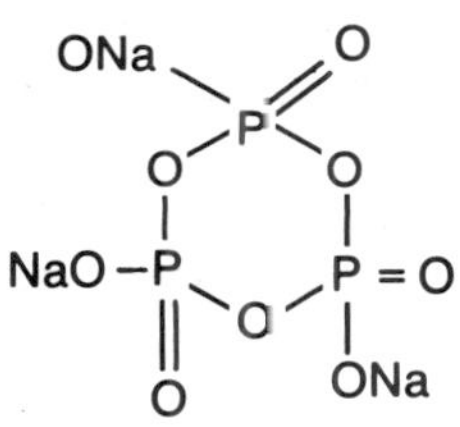

Sodium trimetaphosphate

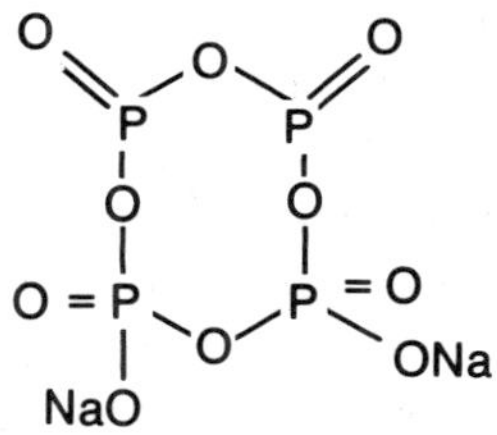

Sodium tetrametaphosphate

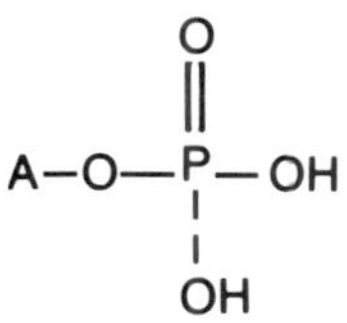

Adenosine monophosphate

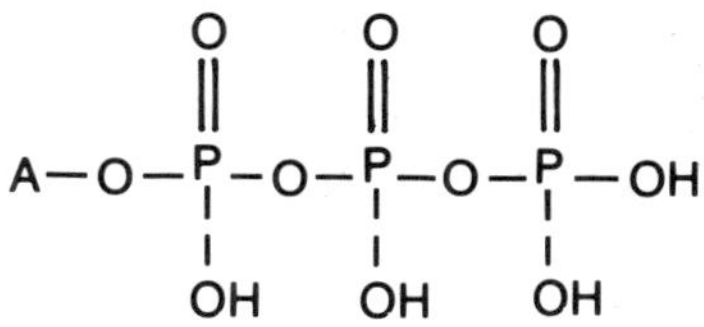

Adenosine triphosphate

Adenosine (represented as A) combined with ortho-, pyro-, and tripolyphosphoric acids to give adenosine monophosphate (AMP), adenosine diphosphate (ADP), and adenosine triphosphate (ATP) molecules which are of great importance in bioenergetics. All of these are excellent chelating agents. An increase in chain increases the the formation constants with Ca and Mg.[1] The structure of the chelates may be represented as follows:

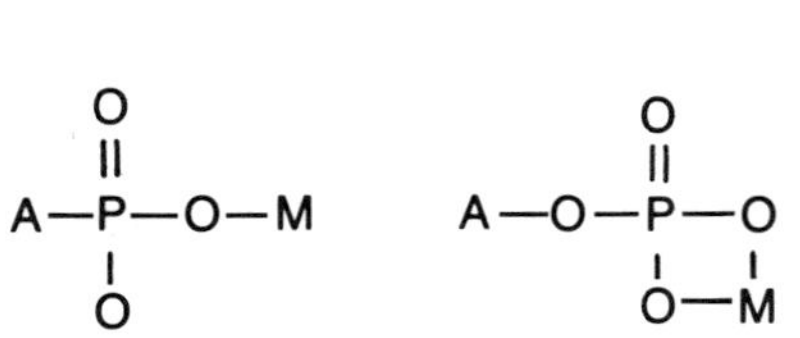

AMP - M Chelates

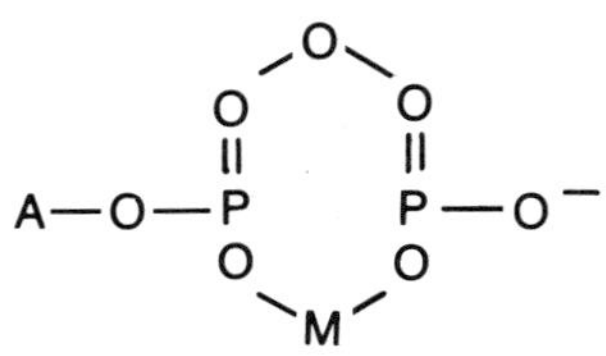

ADP - chelate

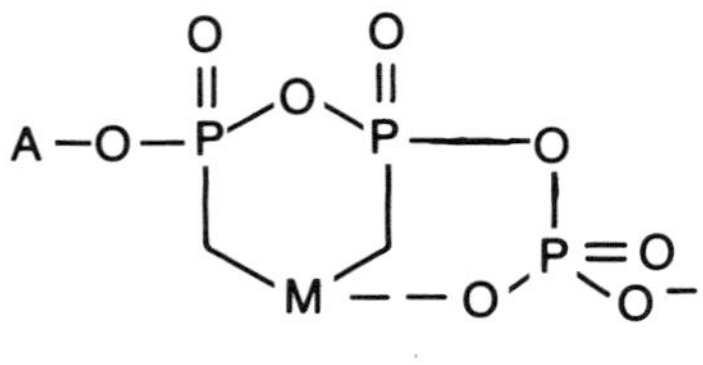

ATP - chelate

II. PHYTATES

In the plant kingdom, phosphates are stored in seeds in combination with inositol or as phytates. Winterstein[2] in 1897 extracted phytic acid from mustard seeds which yielded inositol and orthophosphoric acid on hydrolysis. This product was named phytic acid. The term phytic acid implies myo-inositol hexaphosphoric acid. Ca-Mg salts of phytic acid are designated as phytin. The term phytate is applied loosely for mono to dodeca anion of phytate along with lower esters than hexaphosphate.[3]

A number of reviews are available on the importance of phytic acid in food science and nutrition.[3-6]

Actually, about 60 to 90% of the P present in cereals and oilseeds is in the form of phytic acid (myo-inositol-1,2,3,4,5,6-hexakis-dihydrogenphosphate), but its structure is still controversial. It contains six phosphate molecules for every one molecule of inositol. The studies of Neuberg[7] indicated that the composition of phytic acid was $C_6H_{24}O_{27}P_6$ with 18 titrable hydrogens, and it contained three P–O–P linkages between adjacent phosphates. This is supported by Posternak and Posternak,[8] Fischler and Kurten,[9] Gosselin and Coghlan,[10] and Brown et al.[11] Anderson[12] proposed the formula as $C_6H_{18}P_6O_{24}$ with 12 titrable hydrogens but without three P–O–P linkages. This structure is supported by Barre et al.,[13] Desjobert and Fluerent,[14] Johnson ant Tate,[15] Blank et al.[16] and Costello et al.[17] Maddaiah et al.[18] and Vohra et al.[19] found 12 titrable hydrogens. It is possible that both of these structures may exist in equilibrium. Hoff-Jorgensen[20] prepared the Na salt of phytic acid and the sample dried at 120°C had the formula of $C_6H_6O_{24}P_6Na{\cdot}3H_2O$. This structure was further confirmed by Vohra et al.[19] Erdman[21] postulates the existence of both structures; however, the consensus is in support of Anderson's structure.

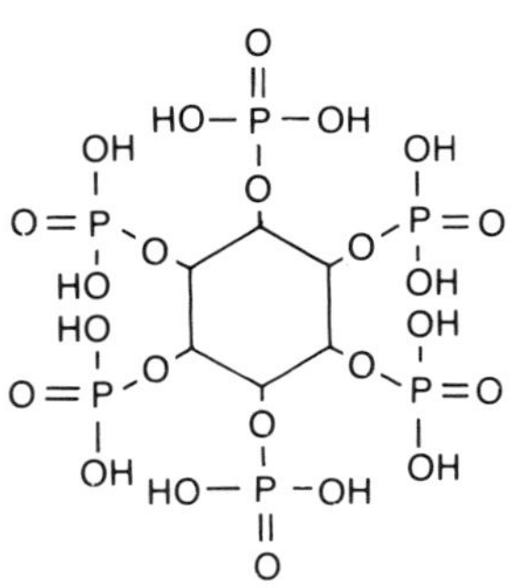

Anderson's structure

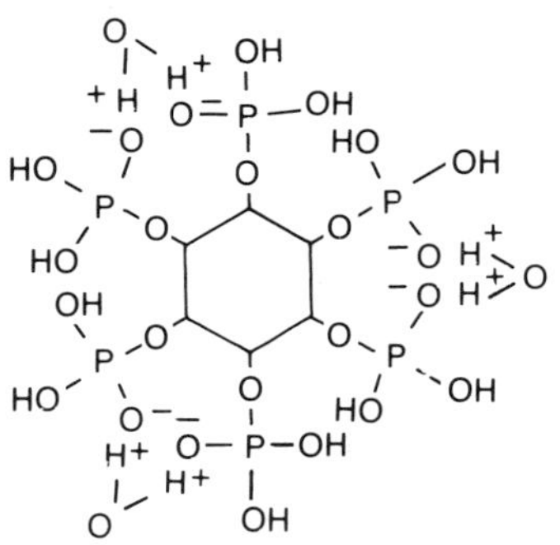

Hoff-Jorgensen's structure

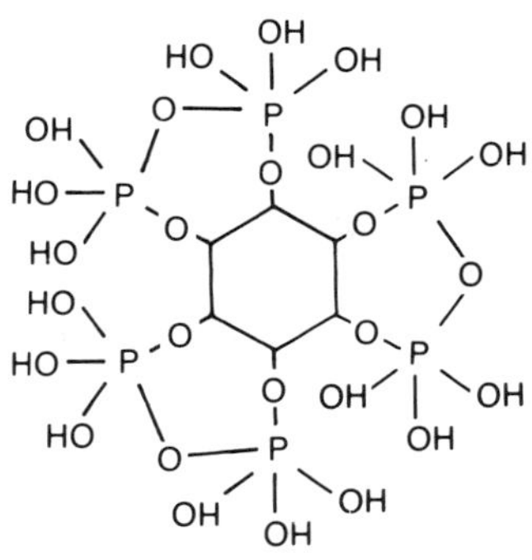

Neuberg's structure

Phytic acid has six reactive phosphates and meets the criterion of a chelating agent. In fact, a cation can complex not only within one phosphate or between two phosphate molecules of the phytic acid, but also between two phytic acid molecules. The structure of such phytates as suggested by Erdman[21] is given below:

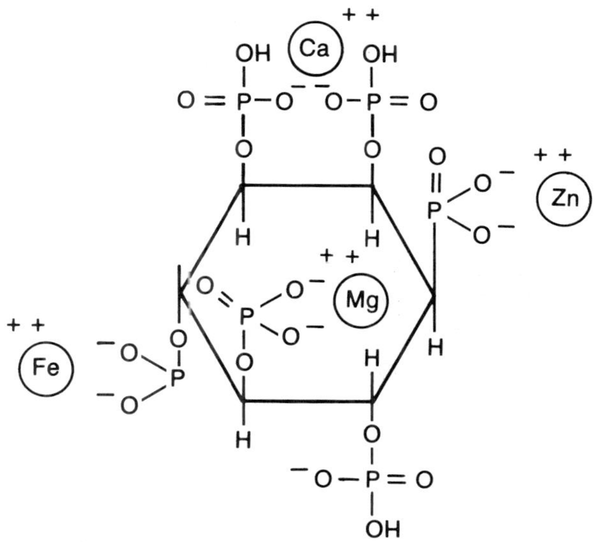

Free phytic acid is an unstable compound and readily releases some orthophosphoric acid. It is generally isolated as a Na or Ca salt.

Most of the phosphorus is stored in seeds in the form of phytic acid derivatives. Minerals are chelated to phytic acid at the storage sites. However, the distribution of phytic acid in cereals and oilseeds is not uniform (Table 1). Many times, phytic acid is associated with some definite component of the seed and may be extracted along with that component. In barley, phytic acid is present in the form of K and Mg salts.[22] About 90% of the total phytic acid in corn is concentrated in the germ rather than in the endosperm and hull portion. Most of the phytate is concentrated in the germ, aleurone layers of the cells, and in the bran or hull in wheat and rice, but not in the endosperm.[23,24]

In dry beans *(Phaseolus vulgaris),* phytic acid is combined with proteins and can be extracted with water.[25]

Phytic acid is distributed throughout the kernel in oilseeds and is concentrated in substructures within the protein body membrane as in peanuts, cottonseed, hempseed, and sunflowers,[26,27] and rapeseed.[28] However, in contrast to other oilseeds, soybeans appear to have no definite site of phytic acid localization,[29] and the concentration of phytic acid does not increase as the protein from soybean is concentrated[30] (Table 1).

The distribution of phytic acid in various parts of the kernel is given in Table 2.

The concentration of phytic acid in oilseeds is given in Table 3.

A. Determination of Phytate

A specific method for the estimation of phytic acid is lacking. Phytic acid is estimated by precipitating it as ferric phytate.[38,39] An equimolar solution of phytic acid and ferric chloride produces a precipitate which is insoluble in the pH range of 1 to 3.5. However, the solubility increases above pH 4 and is about 50% at pH 10. In the presence of an excess of ferric chloride, ferric phytate solubility reaches 90% at pH 1.5, but decreases at higher pH values. Ferric chloride precipitates some inorganic phosphate as well as some other inositol polyphosphates and is not very specific for phytic acid.

Table 1
PHYTIC ACID CONTENT OF SOME CEREALS AND OILSEEDS

Sample	% Phytic acid	Ref.
Barley	0.97—1.08	31, 33
Corn	0.89	21
Corn germ	6.4	21
Oats	0.84—1.01	31, 33
Rice, brown	0.89	24
Rice, wild	2.2	33
Sorghum, low tannin	0.57	34
Sorghum, high tannin	0.97	34
Wheat	1.1	6
Wheat bran	4.8	32
Lima beans	2.52	33
Navy and pinto beans	0.54—1.58	25
Peas, dehydrated	0.9	32
Cottonseed flour, glandless	4.8	21
Peanuts	1.9	32
Sesame seeds, dehulled	5.3	32
Soybeans	1.4	21

Table 2
DISTRIBUTION OF PHYTIC ACID IN CORN, WHEAT, AND RICE KERNELS[23]

	Corn			
	Normal	High-lysine	Wheat	Rice
In kernel,				
Total P (%)	0.30	0.31	0.42	0.31
Phytate P (%)	0.25	0.27	0.32	0.25
	% Distribution			
Germ	88.0	88.9	12.9	7.6
Endosperm	3.2	3.0	2.2	1.2
Aleurone			87.1	
Hull	0.4	1.5	0	
Pericarp				80.0

Harland and Oberleas[33] extracted the phytic acid with 1.2% HCl and ran the solution through an ion-exchange column to remove inorganic phosphates. After recovery of phytate, its P content was determined to calculate the phytic acid content. The method eliminated the precipitation of ferric phytate. Uppstrom and Svensson[37] extracted phytate with 15% trichloroacetic acid and hydrolyzed it with phytase for 16 hr at room temperature to release phosphoric acid. The phosphoric acid was estimated spectrophotometrically. Latta and Eskin[40] used 2.4% HCl for 1 hr for phytate extraction and purified it on an anion-exchanger. The final color was developed with $FeCl_3$ and sulfosalicyclic acid and measured spectrophotometrically. High performance liquid chromatography has also been used in the determination of phytates from beans, oilseeds, and cereal products.[32]

B. Solubility of Phytates

Cereal and oilseed phytates appear to be quite soluble in water. An extraction of 1 part

Table 3
PHYTIC ACID CONCENTRATION IN VARIOUS OILSEED PRODUCTS

Sample	% Phytic acid	Ref.
Cottonseed flour, glanded, defatted	2.86—2.9	21, 35
Cottonseed flour, glandless, defatted	4.29—4.8	21, 35
Cottonseed globoids	60.0	21
Cottonseed flour, air-classified, glandless	3.35	35
Cottonseed, toasted kernel	2.49	35
Peanut meal, defatted, dehulled	1.7	21
Rapeseed protein, concentrated	5.3—7.5	21
Rapeseed meal	3—5	36, 37
Sesame seed	5.18	24
Sesame meal, dehulled	3.6	21
Sesame meal, defatted, dehulled	5.2	21
Soybean meal	1.4—1.6	21
Soybeans	1.0—1.47	31
Soybean protein, isolated	1.6—2.2	21

Table 4
APPARENT FORMATION CONSTANTS OF Ca COMPLEXES (IONIC STRENGTH 0.15 *M*, pH 7.4, TEMP. 37°C)[10]

Ligand	Mole ratio (n)	$-\log K_c$
Hexametaphosphate	2	7.62
Tripolyphosphate	1	3.84
Phytate	1	3.16
Pyrophosphate	1	3.17
Tetrametaphosphate	1	2.76
Trimetaphosphate	1	2.02
Orthophosphate	1	1.79

of beans *(Phaseolus vulgaris)* with 10 parts of water removed 99.6% of phytates.[25] Water extracted 97, 69, 0, 86, and 13% of phytate from soybean flakes, soybean meal, isolated soybean protein, corn germ, and sesame meal, respectively.[24] Again, in water, phytate from corn germ was more soluble, from soybean flakes proportionally soluble, and from sesame meal less soluble than total protein.[30] Two protein-rich fractions have been isolated from peanuts and one contains 0.5% and the other 5.7% phytic acid.[41] Protein concentrates from rapeseed contain over 7% phytic acid.[42]

Phytate appears to exist in most seeds as K-Mg phytate (K_2Mg_5 phytate) and not as phytin (Ca_5Mg phytate).[30]

III. METAL COMPLEXES

Jackman and Black[43] found that Mg and Ca salts of phytic acid tended to be more soluble at lower than at higher pH values. The pH value below which the solubilities increase is about 5.5 for Ca and 7.2 for Mg. Zn phytate was insoluble above pH 4.3. The problem of the solubility of ferric phytate has already been discussed.

Gosselin and Coghlan[10] used the ion-exchange method to determine the apparent formation constants of Ca complexes with some phosphates and their results are presented in Table 4.

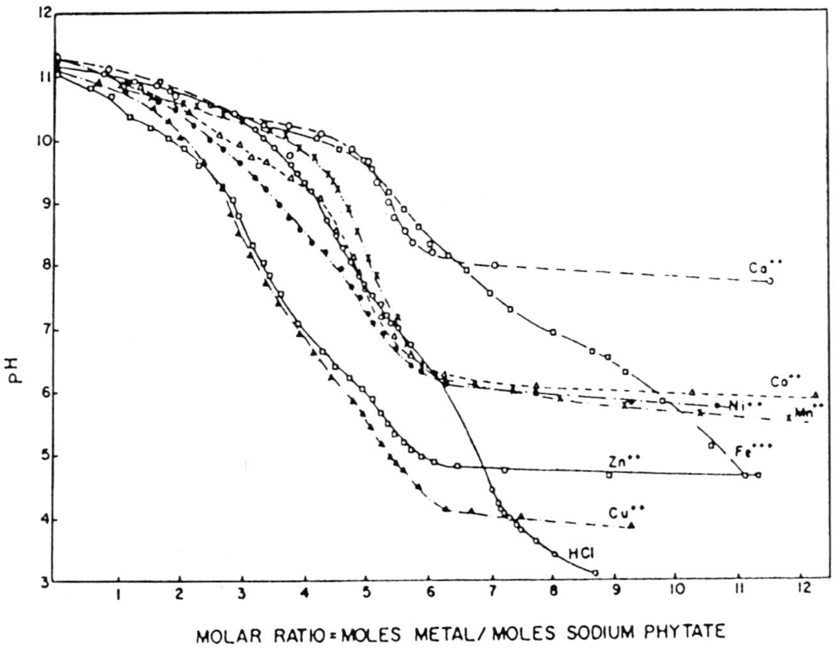

FIGURE 1. Titration of sodium phytate with various metal and hydrogen ions. (From Vohra, P. et al., *Proc. Soc. Exp. Biol. Med.*, 120, 447, 1965. With permission.)

The apparent formation constants were influenced by the ligand to P ratios. Hexametaphosphate and tripolyphosphate complex Ca more strongly than phytate under these conditions. Also, the existance of phytate in a polymeric form is suggested.

Maddaiah et al.[18] and Vohra et al.[19] used a pH drop method to titrate phytate against various metal ions. The data of Vohra et al.[19] of titration of sodium phytate against standard HCl and against metal solutions are presented in Figure 1.

Sodium phytate was assigned a molecular weight of 977.8 corresponding to Anderson's formula with three molecules of water. In the pH zone of 11.2 to 7.5, about 5.0 equivalents of HCl were titrated per equivalent of sodium phytate. Another 2.25 mol were titrated between the pH zone of 7 to 4. The titrations corresponded to the following stepwise reactions:

$$Na_{12}Phytate + 5\ HCl = Na_7H_6Phytate + 5\ NaCl$$

$$Na_7H_6Phytate + 2\ HCl = Na_6H_7Phytate + 2\ NaCl$$

$$Na_6H_7Phytate + 5\ HCl = H_{12}Phytate + 5\ NaCl$$

If the solubility product of the metal phytate which is precipitated is small, the change in phytate ion (indicated by pH) at the equivalence point is large. A sharp drop in the pH of the solution is an indication of a high stability of the metal-phytate complex. The capacity of sodium phytate to form a complex with a metal is dependent on the pH of the solution. At a pH of 7.4, sodium phytate binds metal in the following decreasing order: Cu^{2+}, Zn^{2+}, Ni^{2+}, Co^{2+}, Mn^{2+}, Fe^{2+}, Ca^{2+}. In each titration, a distinct break in the curve takes place when the ratio between the moles of metal to moles of sodium phytate phytate is about five.

Davis et al.[44] washed isolated soybean portein with water and found that the metal ions were reduced to the following levels: Ca, 12%,; Mg, 6%; Mn, 25%; Cu, 116%; Fe, 115%; Mo, 60%; Zn, 63%. The implication is that water did not extract any Cu and Fe, and as

Table 5
pH VALUES FOR PRECIPITATION OF METAL-PHYTATE COMPLEXES FROM SODIUM PHYTATE AND PHYTIC ACID SOLUTIONS[19]

	pH values for precipitation from	
	Sodium phytate	Phytic acid
Ca^{2+}	10.4	?
Fe^{3+}	7.33	From start
Mn^{2+}	10.03	3.9
Co^{2+}	9.4	4.4
Cu^{2+}	9.1	3.0
Ni^{2+}	9.0	5.2
Zn^{2+}	6.3	3.2

the water-soluble fraction disappeared, their concentration in the residue increased. If the protein was washed with Na_2EDTA solution, the residual minerals were as follows: Ca, 8%; Mg, 3%; Mn, 38%; Cu, 53%; Fe, 48%; Mo, 30%; Zn, 4%. EDTA removed more Zn, Fe, and Cu than water, and the protein became practically free of Zn.

Actually, the order of formation constants as given above is not strictly applicable in availability of these metal ions to animals. A more realistic order for decreasing stability may be as follows: Cu, Fe, Zn, Ni, Co, Mn, Ca. (See Table 5.)

The complex formation between metals and freshly prepared phytic acid was also studied. A slight excess of metal ion was added to phytic acid before it was titrated against standard NaOH. The results were in agreement with the earlier results. However, the titrations were complicated by the fact that metal phytates were insoluble and it was difficult to differentiate when metal hydroxides started precipitating along with phytates.

Ion-exchange method could not be used to determine formation constants for Zn-phytate complexes at a pH of 7.4 because of their precipitation.

IV. SOLUBILITY OF MIXED SALTS

Animal studies have indicated that the availability of Zn was significantly reduced on diets containing isolated soybean protein and this is attributed to the presence of phytic acid.[45,46] The symptoms of Zn deficiency are further aggravated by an increase in the Ca level of the diet, and an interaction between Ca, Zn, and phytate is postulated. From their in vitro experiments, Byrd and Matrone[47] demonstrated that at wide Zn to Ca ratios, Ca enhanced the incorporation or adsorption of Zn into phytate. This was further confirmed by Oberleas et al.,[48] who concluded that the formation of Ca-Zn phytate complexes was the mechanism whereby Zn was made less available and the more complete precipitation of Zn in the presence of Ca explains the effect of excess Ca. Oberleas and Moody[49] have further extended these observations to conclude that phytate might affect the bioavailability of Zn, Mn, Cu, Mg, and Ca, but particularly of the trace elements. The precipitation of metal phytates is greatly influenced by their relative concentrations and pH.[18,19,49]

The values of formation constants determined by titrations have no relevence in predicting the power of various chelating agents to improve the availability of Zn from isolated soybean

Table 6
SOLUBILIZING EFFECT OF VARIOUS LIGANDS[52]

% of acid extract pptd. on raising pH to 7

	Zn	Fe	Ca	Mg	Cu
EDTA	0	0	0	19	0
Citric acid	6.9	0	37	16	0
Histidine	51	62	97	49	0
Ascorbic acid	98	71	87	45	0
Lactose	99	78	83	44	0
HCO_3	99.5		75	83	0

protein.[50] Similarly, it is difficult to draw definite conclusions from in vitro experiments because Zn and Mn are also removed to a large extent from solutions by sodium phytate, trisodium phosphate, or dipotassium hydrogen phosphate when the molar ratios of Ca to Zn of Ca to Mn were about 60:1.[88] In fact, Zn was poorly available from its complexes with orthophosphoric acid, hexametaphosphoric acid, pyrophosphoric acid, tripolyphosphoric acid, and phytic acid for poults[50] and from Zn-phosphate for rats.[51]

Lyon[52] conducted in vitro studies on a number of breakfast cereals containing 0.03 to 2.59% phytate on the potential release and fate of Ca, Mg, Cu, and Zn from these cereals through the stomach and small intestine. The phytate to Zn molar ratios ranged from 5.9 to 43.6. A solution of HCl (pH 1) extracted Zn, Ca, and Mg more or less completely from all the cereals except Muesli over 0.5 hr. About 50% of Zn was immediately extracted from Muesli by HCl, but was readsorbed over time and only 18% was extracted finally. About 16% Ca was mobilized immediately, but only 6% was extracted because most of it got readsorbed. The precipitation of minerals with bicarbonate was as follows and Cu remained in solution:

$$Zn^{2+} > Fe^{2+} > Ca^{2+} > Mg^{2+} > Cu^{2+}$$

The importance of solubilizing effect of other ligands is indicated in Table 6.

This further confirms that data as developed by Vohra et al.[19] in model systems giving the following sequence are not always applicable in presence of multiligands:

$$Cu^{2+} > Zn^{2+} > Co^{2+} > Mn^{2+} > Fe^{2+} > Ca^{2+}$$

V. ROLE OF PHYTASE

An enzyme phytase can hydrolyze phytate to release phosphate molecules and make them available to supply the P requirements of the animal. Hydrolysis of phytate by phytase into partially phosphorylated inositols is a stepwise process. Microorganisms easily break down phytate to release P and the bound materials. For this reason, the availability of phytate-bound minderals is greater in ruminants than in monogastric animals. Germ-free chicks[53] and germ-free rats[54] are unable to hydrolyze phytates. The studies with conventional rats suggest that intestinal phytase and phosphatase do not play a role in the adaptive increase in phytate digestibility on marginal-P diets. The adaptation may result from an enhanced phytase or alkaline phosphatase synthesis by intestinal microflora under these conditions.[55]

VI. PHYTATE-PROTEIN INTERACTION

Animals fail to show a Zn deficiency if the protein in their diet is supplied by soybean meal rather than by isolated soybean protein. This has been attributed to the presence of some protein-phytic acid complexes. Allred et al.[46] found that less Zn was bound by an isolated soybean protein from which a part of phytic acid had been removed than the original protein. The amount of Zn bound by an isolated soybean protein and by alpha-casein increased in the presence of sodium phytate, especially at pH values below 4.5.

Formation of phytic acid complexes with proteins is well established and the mechanism of the reaction is pH dependent. According to Cheryan,[5] a unionized protein-phytate complex is probably formed below the isoelectric point and is soluble at $<$ pH 3. A ternary protein-phytate complex with the binding of a cation to protein at the histidine site is formed at pH $>$ 6. The stability of this complex increases up to pH 10. This compound dissociates at pH $>$ 10, resulting in poor solubility of phytic acid but better solubility of protein. The possible structure of the phytate-protein ternary complex is postulated to be as follows where M is the metal:

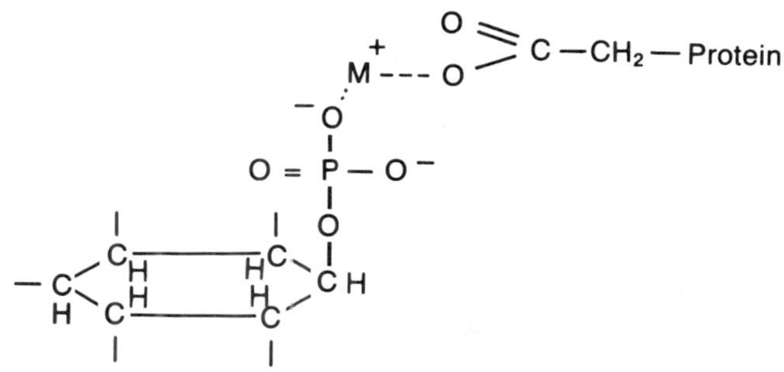

McKinney and Solars[56] isolated a soybean protein containing 1.7 to 3.0% phytate from defatted soybean flakes by extraction with sulfurous acid at pH 2.3 and precipitation at pH 4.2. The phytic acid could not be dialyzed out of this protein. Smith and Rackis[57] had also shown that during the extraction of a protein, it formed a complex with phytic acid. Such complexes were resistant to pepsin digestion.[13] Addition of sodium phytate equivalent to 2% phytic acid to raw wheat starch reduced its digestion by human saliva in vitro by 50% and the inhibition was overcome by the addition of Ca, which complexes with phytic acid.[58]

The topic of soybean protein-mineral interrelationship involving phytic acid was reviewed by Kratzer.[59] No general statements can be made about the native state of phytic acid in plant seeds. It may be complexed with a protein and in other case with divalent cations, or with proteins and divalent cations.[30] The implication is that different isolated soybean proteins do not contain the same amount of phytic acid and are not equivalent in reducing the availability of trace elements. Vohra and Kratzer[60] pointed out that different sources of isolated soybean protein differed markedly in the availability of Zn present in them.

VII. NUTRITIONAL IMPLICATIONS OF PHYTIC ACID

Phytic acid is largely blamed for complexing dietary essential minerals in cereals and oilseeds and render them poorly available to monogastric animals.[21,23] Phytate-protein complexes also play some role in this.[59] The availability of minerals is improved by autoclaving the protein[61,62] or by supplementation with EDTA and some other chelating agents.[63-66]

Unfortunately, these theories do not explain all the observations on trace element availability. This can be better visualized by considering the effect of phytic acid on the availability of specific minerals. However, the information developed on different species of laboratory animals may not always be in agreement. Taylor and Coleman[67] found differences in the absorption of Ca and the availability of phytate-P between rats and hamsters. Rats or chicks may not be the best animal model to predict the effect on humans.

A. Ca

Phytic acid was associated with anticalcifying and rachitogenic properties of cereals. An addition of phytic acid to the diet of dogs reduced Ca absorption and caused rickets.[68] Van Den Berg et al.[69] found phytate to be inert but phytic acid and its hydrolysis products were effective in causing rickets in rats.

Actually, two Ca-phytates can be formed depending upon the molar ratios of phytate and Ca. As the ratio of phytate to Ca is increased, an increasing proportion of Ca becomes unavailable and more phytate becomes hydrolyzable by phytase. Vitamin D increases the availability of phytate-P. The rachitogenicity of high-phytate, low-Ca diets was reduced by Vitamin D. Walker[70] also pointed out that the incrimination of phytate in the etiology of rickets was far from conclusive as an interaction existed between phytate, Ca, and vitamin D.

Walker et al.[71] reported an improvement in Ca and Mg retention in humans as a gradual adaptation took place to cope with a high dietary phytate level after a short period. The human adaptation to high dietary phytate was not confirmed by Reinhold et al.[72] Also, Ca absorption was due to fiber rather than the phytate content of the cereals.[73]

B. Mg

An addition of sodium phytate to casein-based diets aggravated the Mg-deficiency symptoms.[74] Mg absorption, but not balance, was reduced by diets based on isolated soybean protein in comparision to diets containing egg white protein.[75]

C. Fe

The data of the role of phytate on Fe availability are conflicting.[76] Davies and Nightingale[77] found that phytate inhibited Fe absorption, but this was ruled out in many other studies.[65,78-79] May et al.[80] isolated monoferric phytate from wheat bran and established its identity by Mossbauer spectroscopy, and Fe from this compound was readily available to rats.[81]

D. Zn

Dietary phytate depresses the absorption of Zn from the GI tract of chickens[45,48] and rats. House et al.[82] found a decrease in the endogenous fecal Zn excretion in rats fed phytate, which is contrary to the finding of Flanagan.[83]

More data are available on the interaction between Zn, Ca, and phytate on Zn availability than on other elements. Actually, Zn is more available from soybean meal than from isolated soybean proteins. However, the isolated soybean proteins differ among themselves.[60,84] This may be a result of different conditions of protein isolation and the type of phytate-protein-mineral complex obtained during isolation.[85-87]

REFERENCES

1. **Chaberek, S. and Martell, A. E.,** *Organic Sequestering Agents,* John Wiley & Sons, New York, 1959.
2. **Winterstein, E.,** Ueber einen phosphorhaltigen Pflanzenbestandtheil welcher bei der Spaltung Inosit liefert, *Ber. Dtsch. Chem. Ges.,* 30, 2299, 1897.
3. **Maga, J. A.,** Phytate: its chemistry, occurrence, food interactions nutritional significance, and methods of analysis, *J. Agric. Food. Chem.,* 30 1, 1982.
4. **Oberleas, D.,** Phytates, in *Toxicants Occurring Naturally in Foods,* National Academy of Sciences, Washington, D.C., 1973, 363.
5. **Cheryan, M.,** Phytic acic interactions in food systems, *CRC Crit. Rev. Food Sci. Nutr.,* 297, 1980.
6. **Graf, E.,** Applications of phytic acid, *J. Am. Oil Chem. Soc.,* 60, 1861, 1983.
7. **Neuberg, C.,** Zur Frage cer Konstitution des ''Phytins'', *Biochem. Z.,* 9, 557, 1980.
8. **Posternak, S. and Posternak, T.,** Sur la configuraion de l'inosite inactive, *Helv. Chim. Acta,* 12, 1165, 1929.
9. **Fischler, F. and Kurten F. H.,** Über einfachere Nachweismethoden von Inosit und Phytinaten sowei über ein definiertes, Kristallisiertes Barium-phytinat, *Biochem. Z.,* 254, 138, 1932.
10. **Gosselin, R. E. and Coghlan, E. R.,** The stability of complexes between calcium and orthophosphate, polymeric phosphate and phytate, *Arch. Biochem. Biophys.,* 45, 301, 1953.
11. **Brown, E. C., Heit, M. L., and Ryan, D. E.,** Phytic acid: an analytical investigation, *Can. J. Chem.,* 39, 1290, 1961.
12. **Anderson, R. J.,** A contribution to the chemistry of phytin, *J. Biol. Chem.,* 17, 171, 1914.
13. **Barre', M. R.,** Influence de l'acide phytique sur la digestion pepsique de differentes proteines, *Ann. Pharm. Fr.,* 14, 182, 1956.
14. **Desjobert, A. and Fluerent, P.,** Influence de la reaction du milieu sur l'hydrolyse chimique de l'inositol hexaphosphate. Considerations sur la constitution de ce derive, *Bull. Soc. Chim. Biol.,* 36, 475, 1954.
15. **Johnson, L. F. and Tate, M. E.,** Structure of phytic acids, *Can. J. Chem.,* 47, 63, 1969.
16. **Blank, G. E., Pletcher, J., and Sax, M.,** The structure of myoinositol hexaphosphate dodecasoidum salt ocatria contahydrate: a single crystal x-ray analysis, *Biochem. Biophys. Res. Commun.,* 44, 319, 1971.
17. **Costello, A. J. R., Glonek, T., and Myers, T. C.,** ^{31}P-Nuclear magnetic resonance — pH titrations of myo-inositol hexaphosphate, *Carbohydr. Res.,* 46, 159, 1976.
18. **Maddaiah, V. T., Kurnick, A., and Reid, B. L.,** Phytic acid studies, *Proc. Soc. Exp. Biol. Med.,* 115, 391, 1964.
19. **Vohra, P., Gray, G. A., and Kratzer, F. H.,** Phytic acid-metal complexes, *Proc. Soc. Exp. Biol. Med.,* 120, 447, 1965.
20. **Hoff-Jorgenson, E.,** Investigations on the solubility of calcium phytate, *Mat Fys Medd.,* 21, 1, 1944.
21. **Erdman, J. W.,** Oilseed phytates: nutritional implications, *J. Am. Oil Chem. Soc.,* 56, 736, 1979.
22. **Pomeranz, Y.,** Structure and mineral composition of cereal aleurone cells as shown by scanning electron microscopy, *Cereal Chem.* 50, 504, 1973.
23. **O'Dell, B. L., de Boland, A. R., and Koirtyohann, S. K.,** Distribution of phytate and nutritionally important elements among the morphological components of cereal grains, *J. Agric. Food Chem.,* 20, 718, 1972.
24. **de Boland, A. R., Garner, G. B., and O'Dell, B. L.,** Identification and properties of ''phytate'' in cereal grains and oilseed products *J. Agric. Food Chem.,* 23, 1186, 1975.
25. **Lolas, G. M. and Markakis, P.,** Phytic acid and other phosphorus compounds of beans, *J. Agric. Food Chem.,* 23, 13, 1975.
26. **Saio, K., Koyama, E., and Watanabe, T.,** Protein-calcium-phytate relationship in soybean. I. Effect of calcium and phosphorus on solubility characteristics of soybean meal protein, *Agric. Biol. Chem.,* 31, 1195, 1967.
27. **Saio, K., Koyama, E., and Watanabe, T.,** Protein-calcium-phytate relationship in soybean. II. Effect of phytic acid on combination of calcium with soybean meal protein, *Agric. Biol. Chem.,* 32, 448, 1968.
28. **Shah, B. G., Jones, J. D., Mc Laughlin, J. M., and Beare-Rogers, J. L.,** Beneficial effect of zinc supplementation in young rats fed protein concentrate from rapeseed and mustard, *Nutr. Rep. Int.,* 15, 1, 1976.
29. **Tombs, M. P.,** Protein bodies of the soybean, *Plant. Physiol.,* 42, 797, 1967.
30. **O'Dell, B. L. and de Boland, A.,** Complexation of phytate with proteins and cations in corn germ and oilseed meals, *J. Agric. Food. Chem.,* 24, 804, 1976.
31. **Lolas, G. M., Palamidis, N., and Markakis, P.,** The phytic acid-total phosphorus in barley, oats, soybeans and wheat, *Cereal Chem.,* 53, 867, 1976.
32. **Graf, E. amd Dintzis, F. R.,** Determination of phytic acid in foods by high-performance liquid chromatography, *J. Agric. Food Chem.,* 30, 1094, 1982.
33. **Harland, B. F. and Oberleas, D.,** A modified method for phytate analysis using an ion-exchange procedure: application to textured vegetable proteins, *Cereal Chem.,* 54, 827, 1977.

34. **Radhakrishnan, M. R. and Sivaprasad, J.,** Tannin content of sorghum varieties and their role in iron bioavailability, *J. Agric. Food Chem.*, 28, 55, 1980.
35. **Wozenski, J. and Woodburn, M.,** Phytic acid (myo-inositol hexaphosphate) and phytase activity in four cottonseed protein products, *Cereal Chem.*, 52, 665, 1975.
36. **Nwokolo, E. N. and Brags, D. B.,** Influence of phytic acid and crude fiber on the availability of minerals from four protein supplements in growing chicks, *Can. J. Anim. Sci.*, 57, 475, 1977.
37. **Uppstrom, B. and Svensson, R.,** Determination of phytic acid in rapeseed meal, *J. Sci. Food. Agric.*, 31, 651, 1980.
38. **Huebner, W. and Stadler, H.,** Über eine Titrationmethode zue Bestimmung des Phytins, *Biochem. Z.*, 64, 422, 1914.
39. **Ellis, R., Morris, E. R., and Constance, P.,** Quantitative determination of phytate in the presence of high inorganic phosphate, *Anal. Biochem.*, 77, 536, 1977.
40. **Latta, M. and Eskin, M.,** A simple and rapid colorimetric method for phytate determination, *J. Agric. Food Chem.*, 28, 1313, 1980.
41. **Dieckert, J. W., Snowden, J. E., Jr., Moore, A. T., Heinzelman, D. C., and Altschul, A. M.,** Composition of some subcellular fractions from seeds of *Arachis hypogaea*, *J. Food Sci.*, 27, 321, 1962.
42. **McLaughlan, J. M., Kones, J. D., Shah, B. G., and Beae-Rogers, J. L.,** Reproduction in rats fed protein concentrate from mustard or rapeseed, *Nutr. Rep. Int.*, 11, 327, 1975.
43. **Jackman, R. H. and Black, C. A.,** Solubility of iron, aluminum, calcium and magnesium inositol phosphates at different pH values, *Soil Sci.*, 72, 179, 1951.
44. **Davis, P. N., Norris, L. C., and Kratzer, F. H.,** Iron deficiency studies in chicks using treated isolated soybean protein diets, *J. Nutr.*, 78, 445, 1962.
45. **O'Dell, B. L. and Savage, J. E.,** Effect of phytic acid on zinc availability, *Proc. Soc. Exp. Biol. Med.*, 103, 304, 1960.
46. **Allred, J. B., Kratzer, F. H., and Porter, J. W. G.,** Some factors affecting the *in vitro* binding of zinc by isolated soybean protein and by α-casein, *Br. J. Nutr.*, 18, 575, 1964.
47. **Byrd, C. A. and Matrone, G.,** Investigation of chemical basis of zinc-calcium-phytate interaction in biological systems, *Proc. Soc. Biol. Med.*, 119, 347, 1965.
48. **Oberleas, D., Muhrer, M. E., and O'Dell, B. L.,** Dietary metal-complexing agents and zinc availability in the rat, *J. Nutr.*, 90, 56, 1966.
49. **Oberleas, D. and Moody, N.,** In vitro interaction of phytate with trace elements, in *Trace Element Metabolism in Man and Animals*, Gawthorne, J. M.,. Howell, J. M. C., and White, C. L., Eds., Springer-Verlag, 1981, 129.
50. **Vohra, P. and Kratzer, F. H.,** Influence of various phosphates and other complexing agents on the availability of zinc for turkey poults, *J. Nutr.*, 89, 106, 1966.
51. **Seal, C. S. and Heaton, F. W.,** Chemical factors affecting the intestinal absorption of zinc in vitro and in vivo, *Br. J. Nutr.*, 50, 317, 1983.
52. **Lyon, D. B.,** Studies on the solubility of Ca, Mg, Zn, and Cu in cereal products, *Am. J. Clin. Nutr.*, 39, 190, 1984.
53. **Savage, J. E., Yohe, J. M., Pickett, E. E., and O'Dell, B. L.,** Zinc metabolism in the growing chick. Tissue concentration and effect of phytate on absorption, *Poult. Sci.*, 43, 420, 1964.
54. **Wise, A. and Gilburt, D. J.,** Phytate hydrolysis by germfree and conventional rats, *Appl. Environ. Microbiol.*, 43, 753, 1982.
55. **Moore, R. J. and Veum, T. L.,** Adaptive increase in phytate digestibility by phosphorus-deprived rats and the relationship of intestinal phytase (*EC* 3.1.3.8) and alkaline phosphatase (*EC* 3.1.3.1) to phytate utilization, *Bri. J. Nutr.*, 49, 145, 1983.
56. **McKinney, L. L. and Sollars, W. F.,** Extraction of soybean protein with sulfurous acid, *Ind. J. Eng. Chem.*, 41, 1058, 1949.
57. **Smith, A. K. and Rackis, J. J.,** Phytin elemination in soybean protein isolation, *J. Am. Chem. Soc.*, 79, 633, 1957.
58. **Yoon, J. H., Thompson, L. U., and Jenkins, D. J. A.,** The effect of phytic acid on in vitro rate of starch digestibility and blood glucose response, *Am. J. Clin. Nutr.*, 38, 835, 1983.
59. **Kratzer, F. H.,** Soybean protein-mineral interrelationship, *Fed. Proc. Fed. Am. Soc. Exp. Biol.*, 24, 1498, 1965.
60. **Vohra, P. and Kratzer, F. H.,** The importance of the source of isolated soybean protein in nutrition, *Poult. Sci.*, 46, 1016, 1967.
61. **Kratzer, F. H., Allred, J. B., Davis, P. N., Marshall, B. J., and Vohra, P.,** The effect of autoclaving soybean proteins and the addition of EDTA on the biological availability of dietary zinc for turkey poults, *J. Nutr.*, 68, 313, 1958.
62. **Kratzer, F. H., Vohra, P., Allred, J. B., and Davis, P. N.,** Effect of zinc upon growth and incidence of perosis in turkey poults, *Proc. Soc. Exp. Biol. Med.*, 98, 205, 1958.

63. **Davis, P. N., Norris, L. C., and Kratzer, F. H.,** Interference of soybean proteins with the utilization of trace minerals, *J. Nutr.*, 77, 217, 1962.
64. **Davis, P. N., Norris, L. C., and Kratzer, F. H.,** Iron utilization and metabolism in the chick, *J. Nutr.*, 94, 407, 1968.
65. **Vohra, P. and Kratzer, F. H.,** Influence of various chelating agents on the availability of zinc, *J. Nutr.*, 82, 249, 1964.
66. **Nielsen, F. H., Sunde, M. L., and Hoekstra, W. G.,** Effect of some dietary synthetic and natural chelating agents on the zinc-deficiency syndrome in the chick, *J. Nutr.*, 89, 35, 1966.
67. **Taylor, T. G. and Coleman, J. W.,** A comparative study of the absorption of calcium and the availability of phytate-phosphorus in the golden hamster *(Mesocricetus auratus)* and the laboratory rat, *Br. J. Nutr.*, 42, 113, 1979.
68. **Mellanby, E.,** The rickets-producing and anti-calcifying action of phytate, *J. Physiol. (London)*, 109, 488, 1949.
69. **Van Den Berg, C. J., Hill, L. F., and Stanbury, S. W.,** Inositol phosphates and phytic acid as inhibitors of biological calcification in the rat, *Clin. Sci.*, 43, 377, 1972.
70. **Walker, A. R. P.,** Cereals, phytic acid and calcification, *Lancet*, 261, 244, 1951.
71. **Walker, A. R. P., Fox, F. W., and Irving, J. T.,** The effect of bread rich in phytate phosphorus on the metabolism of certain mineral salts with special reference to calcium, *Biochem. J.*, 42, 452, 1948.
72. **Reinhold, J. G., Lahingarzadeh, A., Khosrow, N., and Hedayati, A.,** Effects of purified phytate and phytate-rich bread upon metabolism of zinc, calcium, phosphorus and nitrogen, *Lancet*, 1, 283, 1973.
73. **Rheinhold, J. G., Ismail-Beigi, F., and Faraji, B.,** Fiber vs. phytate as determinant of the availability of calcium, zinc, and iron of breadstuffs, *Nutr. Rep. Int.*, 12, 75, 1975.
74. **Roberts, A. H. and Yudkin, J.,** Dietary phytate as a possible cause of magnesium deficiency, *Nature (London)*, 185, 823, 1960.
75. **Forbes, R. M.,** Mineral utilization in the rat. III. Effects of calcium, phosphorus, lactose and source of protein in zinc deficient and zinc adequate diets, *J. Nutr.*, 83, 225, 1964.
76. **Fairweather-Tait, S. J.,** The effect of different levels of wheat bran on iron absorption in rats from bread containing similar amounts of phytate, *Br. J. Nutr.*, 47, 247, 1982.
77. **Davies, N. T. and Nightingale, R.,** The effects of phytate on intestinal absorption and secretion of zinc and whole-body retention of zinc, copper, iron and manganese in rats, *Br. J. Nutr.*, 34, 243, 1975.
78. **Cowan, J. W., Esfahemi, M., Salji, J. P., and Azzain, S. A.,** Effect of phytate on iron absorption in the rat, *J. Nutr.*, 90, 423, 1966.
79. **Morris, E. R. and Ellis, R.,** Isolation of monoferric phytate from wheat bran and its biological value as an iron source to the rat, *J. Nutr.*, 106, 753, 1976.
80. **May, L., Morris, E. R., and Ellis, R.,** Chemical identity of iron in wheat by Mossbauer spectroscopy, *J. Agric. Food Chem.*, 28, 1004, 1980.
81. **Lipschitz, D. A., Simpson, K. M., Cook, J. D., and Morris, E. B.,** Absorption of monoferric phytate by dogs, *J. Nutr.*, 109, 1154, 1979.
82. **House, W. A., Welch, R. M., and Van Campen, D. R.,** Effect of phytic acid on the absorption, distribution, and endogenous excretion of zinc in rats, *J. Nutr.*, 112, 941, 1982.
83. **Flanagan, P. R.,** A model to produce pure deficiency in rats and its use to demonstrate that dietary phytate increases the excretion of endogenous zinc, *J. Nutr.*, 114, 493, 1984.
84. **Rackis, J. J. and Anderson, R. L.,** Mineral availability in soy protein products, *Food Prod. Dev.*, 11 (10), 38, 1977.
85. **Forbes, R. M., Erdman, J. W., Jr., Parker, H. M., Kondo, H., and Ketelsen, S. M.,** Bioavailability of zinc in coagulated soy protein (Tofu) to rats and effect of dietary calcium at a constant phytate:zinc ratio, *J. Nutr.*, 113, 205, 1983.
86. **Davies, N. T. and Reid, H.,** An evaluation of the phytate, zinc, copper, iron and manganese contents of, and Zn availability from soya-based textured-vegetable-protein meat-substitutes or meat-extenders, *Br. J. Nutr.*, 41, 579, 1979.
87. **Ketelson, S. M. Stuart, M. A., Weaver, C. M., Forbes, R. M., and Erdman, J. W., Jr.,** Bioavailability of zinc to rats from defatted soy flour, acid-precipitated soy concentrate and neutralized soy concentrate as determined by intrinsic and extrinsic labeling techniques, *J. Nutr.*, 114, 536, 1984.
88. **Vohra, P. and Davis, M. J.,** Unpublished data.

Chapter 5

OTHER CHELATES

I. OXALIC ACID

Oxalic acid (HOOC–COOH) has two adjoining carboxylic groups and the values for its dissocation constants ($K_1 = 6.5 \times 10^{-2}$ and $K_2 = 6.1 \times 10^{-5}$) are greater than for dicarboxylic acids with longer carbon chains. It is a chelating agent and easily forms oxalates with metal ions. In water, oxalates with Na and K are very soluble, with Mg poorly soluble, and Ca practically insoluble.

Oxalic acid is widely distributed in plants and animals, but plants contain more of it than the animals. The name oxalic acid is derived from the plant *Oxalis* (wood sorrel) from which it was first isolated. It is present as an acid K salt in *Oxalis* and *Rumex* familes, as Na salt in *Salicornia* and *Salsola,* and as Mg salt in *Gramineae* leaves.[1] The oxalic acid content of some foods is given in Table 1.

Oxalic acid is produced in the animal body endogenously. It can be produced from glycine as follows, but appears to be excreted in the urine in mammals:

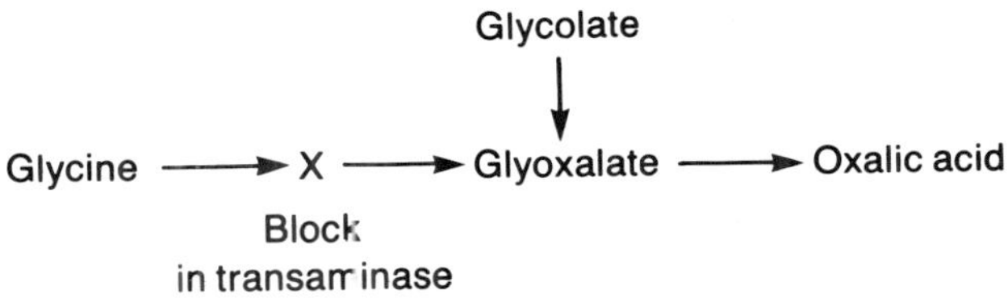

Oxalic acid can also be produced from ascorbic acid.

The information on the toxic effects of oxalic acid has been reviewed by Fassett,[1] Gontzea and Sutzescu,[2] and Oke.[3] A dose of 5 g oxalic acid can be fatal for humans. The toxicity can be a result of a combination with Ca to form insoluble oxalate and create a deficiency of Ca in the body, and an accumulation of oxalic acid itself in the body. However, it is very difficult to establish a direct relationship between the high levels of oxalic acid in a food and its toxicity to humans in the presence of adequate levels of Ca and Vitamin D. It is doubtful that any danger is encountered in consuming foods such as spinach and rhubarb, both rich in oxalic acid. Some cases of death have been reported following consumption of high amounts of rhubarb, but Fassett[1] suspects these to be due to toxic anthraquinone glycosides and not to oxalates.

The absorption of oxalates from fasted animals in much higher than from oxalate mixed with food. The importance of a diet balanced in Ca and vitamin D (to protect animals from oxalate toxicity) cannot be overstressed. If the diet is deficient in Ca, difficulties may arise because of a further reduction in its availability by oxalic acid.

The role of dietary fiber in the presence of oxalates on mineral availability also cannot be ruled out. An increase in fecal and a decrease in urinary oxalic acid was observed when a high-fiber diet was fed to humans as compared to a low-fiber diet. Ca, Mg, and Zn balances were negative on a high-fiber, high-spinach diet.[4]

II. IONOPHORES

Ionophores or ion bearers are naturally occurring compounds which carry or transport small cations selectively across the lipid barriers of the cell membrane. Ionophores (200 to

Table 1
THE OXALIC ACID CONTENT OF SOME FRESH FOODS[2]

Plant	Oxalic acid (%)	Ca (%)	Milliequivalents oxalic acid/Ca
Rhubarb *(Rehum rhaponticum)*	0.27—1.34	0.045	8.5
Common sorrel *(Rumex acetosa)*	0.27—0.73	0.04	5.6
Garden sorrel *(Rumex patientia)*	0.3—0.7	0.045	5.0
Spinach *(Spinacia oleracea)*	0.32—1.2	0.1	4.3
Red beetroots *(Beta vulgaris)*	0.12—0.45	0.024	5.1
Tea *(Thea chinesis)*	0.3—2.0	0.45	1.1
Coffee *(Coffea arabica)*	0.05—0.15	0.012	3.9

2000 mol wt) may be analogous to some component of the biological membrane as they impart selective permeability to artificial membranes.[5]

Valinomycin ($C_{54}H_{90}N_6O_{18}$) was the first ionophore to be isolated by Brockmann[6] from *Streptomyces fulvissimus,* and is composed of 3 mol each of L-valine, D-alpha-hydroxy-isovaleric acid, D-valine, and L-lactic acid linked alternately to form a 36-membered ring. The structure is as follows:

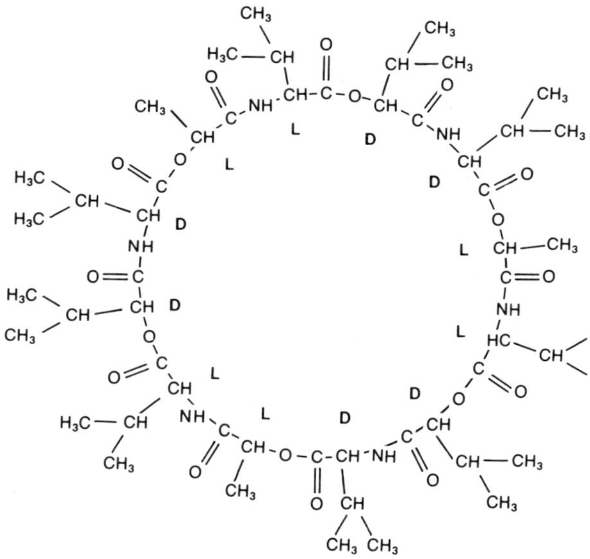

About 75 ionophores have been isolated from various molds and fungi. The ionophores may act as antibiotics. They increase the transport of cations selectively into mitochondria. The antibiotic activity may be a result of their disrupting energy metabolism and ionic equilibrium, causing the death of the microorganism.

Most ionophores can be divided into two groups: neutral ionophores and carboxylic ionophores containing a single-terminal carboxyl group. The neutral ionophores show the following preference for cations:

$$K^+ = Rb^+ > Cs^+ > Na^+ > Li^+$$

The carboxylic ionophores release K^+ in exchange for H^+ in mitochondria and the preference for cations is more variable. The structures of a few of these ionophores are given below.

Lasalocid (Bovatec, Avatec) ($C_{34}H_{54}O_8$) from *Streptomyces*.

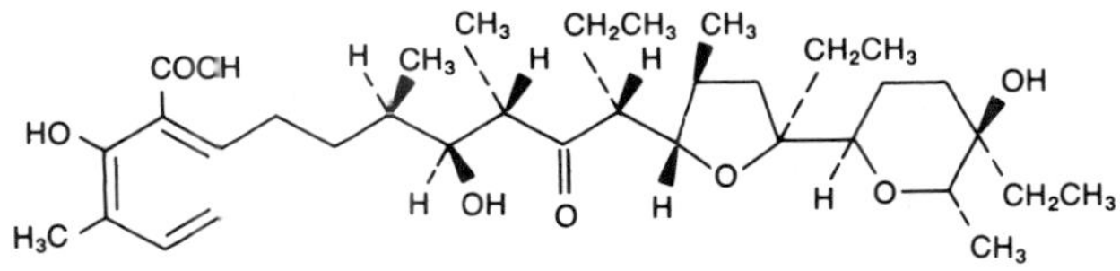

Enniatins from *Fusaria:* Enniatin A ($C_{36}H_{63}N_3O_9$) Enniatin B ($C_{33}H_{57}N_3O_9$).

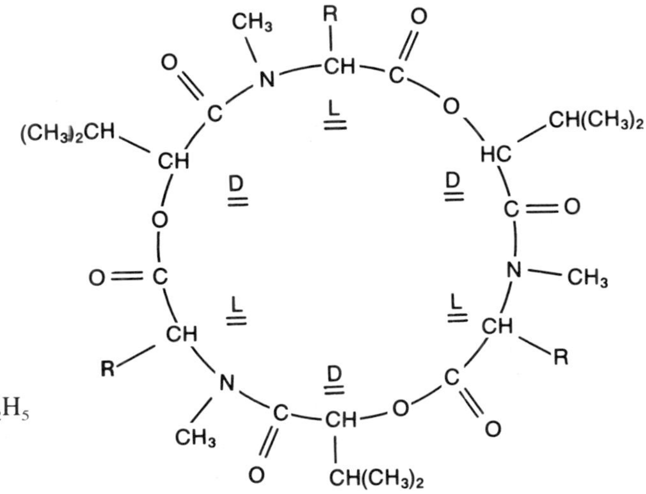

Enniatin A, R = $CH(CH_3)C_2H_5$
Enniatin B, R = $CH(CH_3)_2$

Gramicidins from *Bacillus brevis* are a family of pentadecapeptides, but Gramicidin S ($C_{60}H_{92}N_{12}O_{10}$) is a cyclic decapeptide.

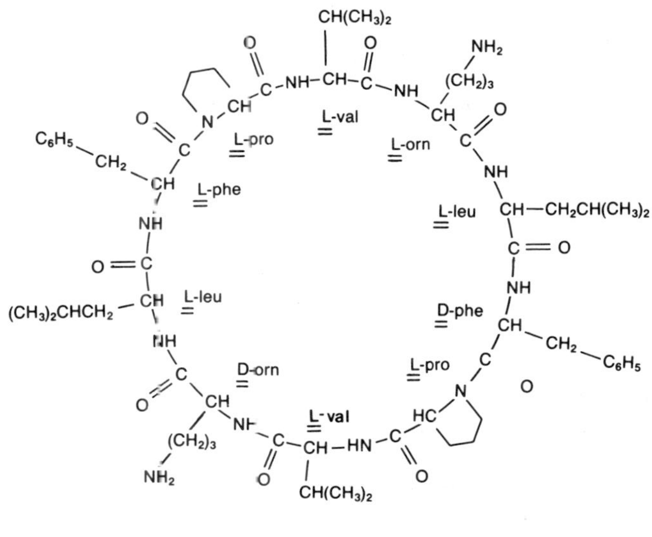

Monensin ($C_{36}H_{62}O_{11}$) from *Streptomyces cinnamonensis*.
Rumensin (Coban) of sodium salt of Monensin.

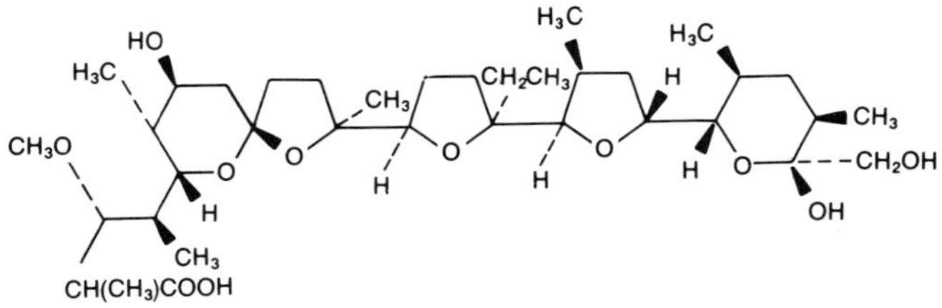

Nigericin ($C_{44}H_{67}NaO_{11}$) from *Streptomyces hygroscopicus* E-749. It resembles monensin with an additional tetrahydropyranol ring.

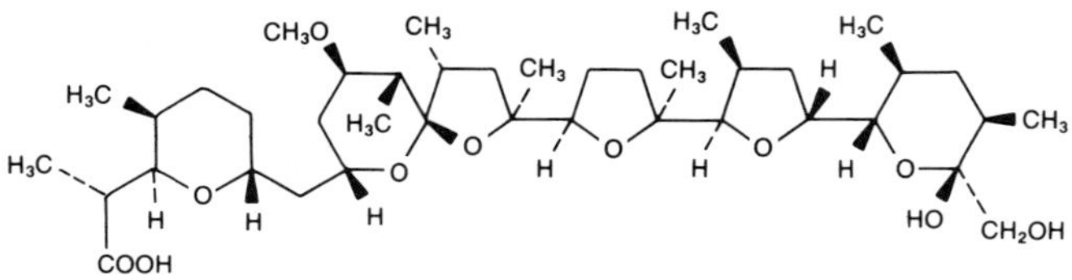

The following carboxylic ionophores are currently used in animal feeds:

- Monesin
- Rumensin
- Lasalocid

Ionophores may be either channel-forming proteins which span the membrane through which cations can pass, or of carrier type. The carrier-ionophore chelate cation prior to its transport across the membrane, is selective in binding of Na^+ or K^+, and Mg^{2+} or Ca^{2+}. The cations are differentiated on the basis of their ionic radii, charge differences, polarization potential, solvation tendency, sphere of hydration of the cation, ligand, and the complex. The stability constants depend upon the size of the ligand. Valinomycin, a cyclic peptide, binds K^+ selectively in preference to Na^+ and transports it into the mitochondria. Nigericin reverses the mitochondrial uptake of K^+.

Monensin preferably binds Na^+ rather than K^+. Rumensin is the Na salt of monensin. As the ability of carboxylic ionophores to cross the poultry digestive tract is limited, monensin is used as a coccidiostat.

Nigericin prefers K^+ over Na^+ by a factor of 100 in contrast to monensin,[7] which prefers Na^+ over K^+ by a factor of 10.

Lasalocid A is selective for Ca^{2+} over Mg^{2+} and is used in the prevention of coccidiosis in sheep. Ionophores appear to improve the growth rates and feed efficiency of ruminants.

III. CLAYS AS CHELATING AGENTS

Geophagia or eating of clays is quite common all over the world since ancient times. This is specially practiced by women during pregnancy. The usual response of the practitioner is a feeling of well-being by geophagia. Specially baked clays are sold in pellets in some parts of Asia and Africa for human consumption. No definite nutritional advantage of geophagia has been documented for humans. However, Danford[8] has questioned the safety of geophagia, as a number of nutritional disorders could result, especially among poorly nourished people. These include nutrient deficiencies of vitamins, trace elements such as Fe and Zn, due to their chelation in the intestinal tract, intestinal perforation, dental injury, and constipation.

Kaolin, a hydrated aluminum silicate, with a formula of $H_2Al_2Si_2O_8 \cdot H_2O$, is specially prepared for medicinal use as an adsorbant in the GI tract, by levigating with water to remove sand. It gives relief against diarrhea and acts as a chelating agent in the intestines.

Bentonites are composed of montmorillonite clay, mostly consisting of $Al_2O_3 \cdot 4SiO_2 \cdot H_2O$. They can absorb ten times their weight of water to form gels or viscous suspensions. They are incorporated in animal feeds to improve pelleting. They have been reported to improve nitrogen utilization in urea- and protein-supplemented diets in ruminants, reduce acidosis, and improve milk production.

Douglas[9] has reviewed the topic of the effect of adding bentonites to ruminant diets. Bentonites can intercollate an enormous range of guest molecules in water or dilute electrolyte solutions. The order of preference of bentonites for monovalent cations is as follows:

$$Cs > Rb > K = NH_4 > Na > Li$$

Proteins are also adsorbed by bentonites, and the adsorption decreases in the order of H > Na > Ca > Al > La > montmorillonite. Amino acids also from stable coordination complexes with bentonites in the presense of chelating cations Cu, Co, or Zn.

Adhesion of montmorillonite to bacteria is increased by the presence of polyvalent cations in the clay. The result is that bentonites modify the reactions of mixed microbial populations. Bentonites are suggested for use in protecting proteins from digestion by rumen microflora.

REFERENCES

1. **Fassett, D. W.,** Oxalates, in *Toxicants Occurring Naturally in Foods,* National Academy of Sciences, Washington, D.C., 1973, 345.
2. **Gontzea, I. and Sutzescu, P.,** *Natural Antinutritive Substances in Foodstuffs and Forages,* S. Karger, Basel, 1968, 84.
3. **Oke, O. L.,** Oxalic acid in plants and in nutrition, in *World Review of Nutrition and Dietetics,* Vol. 10, S. Karger, Basel, 1969, 262.
4. **Kelsay, J. L. and Prather, E. S.,** Mineral balances of human subjects consuming spinach in a low fiber diet and in a diet containing fruits and vegetables, *Am. J. Clin. Nutr.,* 38, 12, 1983.
5. **Brockmann, H. and Schmidt-Kastner, G.,** Valinomycin J. XXVII. Mitteil Über antibiotica aus Actinomyceten, *Chem. Ber.,* 88, 57, 1955.
6. **Pressman, B. C.,** Alkali metal chelators — the ionophores, in *Inorganic Biochemistry,* Vol. 1, Eichhorn, G. L., Ed., Elsevier, Amsterdam, 1973, 203.
7. **Pressman, B. C., Painter, G., and Fahim, M.,** Molecular and biological properties of ionophores, in *Inorganic Chemistry in Biology and Medicine,* Martell, A. E., Ed., ACS Symp. Ser. 140, American Chemical Society, Washington, D.C., 1980, 3.
8. **Danford, D. E.,** Pica and nutrition, *Ann. Rev. Nutr.,* 2, 303, 1982.
9. **Douglas, L.,** Sodium Bentonite as an Additive for Ruminant Diets, Ph.D. thesis, University of New England, Armidale, New South Wales, Australia, 1984.

Chapter 6

ZINC

I. INTRODUCTION

The mechanism of Zn utilization has been reviewed by Cousins,[1] Kirchgessner and Weigand,[2] and Solomons.[3] The efficiency of Zn absorption appears to be homeostatically regulated since Zn is used more efficiently at suboptimal levels and absorption is reduced when excessive levels of Zn are fed. The absorption of Zn is also increased during pregnancy and lactation.

There are four distinct phases of Zn absorption, as outlined by Cousins[1] (see Figure 2 in chapter 3), all of which may exercise some control in the overall utilization of Zn. These are

1. The uptake of Zn by the intestinal cells
2. The intracellular control of Zn absorption
3. The transfer from intestinal cells to the portal circulation
4. The secretion of Zn into the intestine

It is probable that the effects of dietary chelating agents on the nutritional availability of Zn are in the GI tract where they influence the uptake of Zn by the intestinal cells.

II. UPTAKE BY THE INTESTINAL CELLS

The location of the absorption site of Zn from the GI tract has been studied by in vivo and in vitro methods. In the rat, Zn can be absorbed throughout the small intestine, but absorption is minimal from the stomach or large intestine. There is evidence that the jejunum is the major site of absorption in the human. Kirchgessner and Weigand[2] found that Zn depletion of rats stimulated the mucosal-to-serosal transfer of Zn in everted sacs most markedly in the distal jejunum and ileum. While there are some conflicting data, a number of experiments show that the absorption process involves active transport which requires energy. The transport of Zn in everted gut segments may be inhibited by 2,4-dinitrophenol as well as lack of oxygen, metabolizable hexose, and Na in the incubation medium.[2] With the usual human diet, the amount of Zn absorbed is far less than the amount needed to saturate the active transport system. Solomons[3] pointed out that the release of Zn from food by the digestion of protein is a prerequisite for absorption and even if the duodenum had a high affinity for Zn, the need for prior digestion would limit the absorption of Zn in that area.

As will be discussed later, Zn in the intestinal tract may be associated with a variety of ligands which may influence the absorption process. Some of these ligands are present in the diet; others are produced by the digestion of food materials along with those which are secreted by the animal. In an isolated rat intestinal perfusion system,[1] the uptake of Zn by the intestinal mucosa could be studied. Methionine and histidine, for example, enhanced Zn uptake while cysteine and arginine had no effect. The effect of high protein in enhancing Zn absorption may be due to its effect on the intestinal lumen environment. A low molecular weight Zn-binding ligand has been isolated from the rat intestine, but whether such a ligand is involved in the absorption process is not clear. The role of such a ligand in physiological secretions is supported by the fact that ligation of the pancreatic duct caused a reduced uptake of Zn in rats. The presence of Zn in combination with a specific ligand does not necessarily mean that this ligand plays a role in absorption, or is physiologically important.

It is probable that competition for binding Zn among the many ligands present in the intestinal tract is enormous and has a large influence of the availability of Zn for transport to, and possibly across, the brush border of the intestinal epithelial cells.

III. INTRACELLULAR EXCHANGE

As shown in Figure 2 of Chapter 3, after Zn passes across the brush border membrane system, it enters a Zn pool in the intestinal cell to be distributed among a number of different ligands. The function of Zn in many enzyme systems is essential for the survival of the cell. It has been shown that when ^{65}Zn was fed, it could be found later in all of the subcellular fractions of cells prepared by scraping the musoca from the muscle of the intestine. DNA and RNA nucelotidyltransferases are also Zn metalloenzymes and participate in the control of the synthesis of thionein, which is available for metallothionein synthesis with Zn, Cu, or Cd. Mammalian metallothioneins have a primary structure of 61 amino acids, of which 25 to 35% are cysteine and bind up to 5 to 7 g-atoms of metal per molecule of metallothionein. Zn and Cu are the predominant metals, but Cd is also bound in cases of environmental contamination. Metallothionein has no enzymatic function and appears to be an intracellular ligand which may be induced by various stimuli. Administration of Zn, Cu, or Cd to a Zn-deficient rat will induce metallothionein synthesis. Several hormonal relationships to metallothionein have been observed (235). Glucocorticoids induce metallothionein gene expression in the presence of either insulin or glucagon. Glucagon administration produces a transient increase in metallothionein-bound Zn without a change in the metallothionein mRNA.

DiSilvestro and Cousins[4] have proposed that metallothionein functions in the intestinal cell to regulate Zn absorption in response to acute changes in Zn status. An increase in dietary Zn increased the mRNA activity in the intestinal cells and enhanced the binding of Zn to protein. Starcher et al.[213] found that mice injected intraperitoneally with low levels of Zn had increased ^{65}Zn absorption and increased intestinal metallothionein. The stress of handling the mice or the administration of actinomycin D before the Zn administration obliterated the response to Zn. They suggested that the proportionality of Zn absorption to the intestinal metallothionein level implied a significant role for metallothionein in Zn absorption. On the other hand, the metallothionein synthesis was lowest in the Zn-deficient mice and DeSilvestro and Cousins[4] suggest that the long-term changes in Zn and Cu absorption are less closely associated with metallothionein synthesis. Metallothionein may be an essential ligand with which Zn must complex during absorption. Only when the Zn load is excessive will the gene be expressed, which results in metallothionein synthesis and temporary retention of Zn within the cell. Metallothionein is also found in the kidney and its presence in kidney cells is important in Zn homeostasis.

Zn, Cu, and Cd all bind to metallothionein. Zn metallothionein is degraded with a $t_{1/2}$ of 18 to 20 hr while the protein of cadmium metallothionein is degraded with a $t_{1/2}$ of 3.7 days. The binding affinity of metallothionein for Cd is greater than for Zn. The Cd is probably retained by rebinding to nascent apoprotein so that the net degradation of Cd metallothionein occurs even more slowly than is indicated from the protein degradation.

IV. TRANSFER OF Zn FROM INTESTINAL CELLS TO PORTAL CIRCULATION

The mechanism is not well understood, but the transfer of Zn to the portal circulation seems directly proportional to the lumen Zn content. Albumin is probably the carrier plasma protein with which Zn is transported from the intestine, although transferrin has also been proposed as a possible carrier.

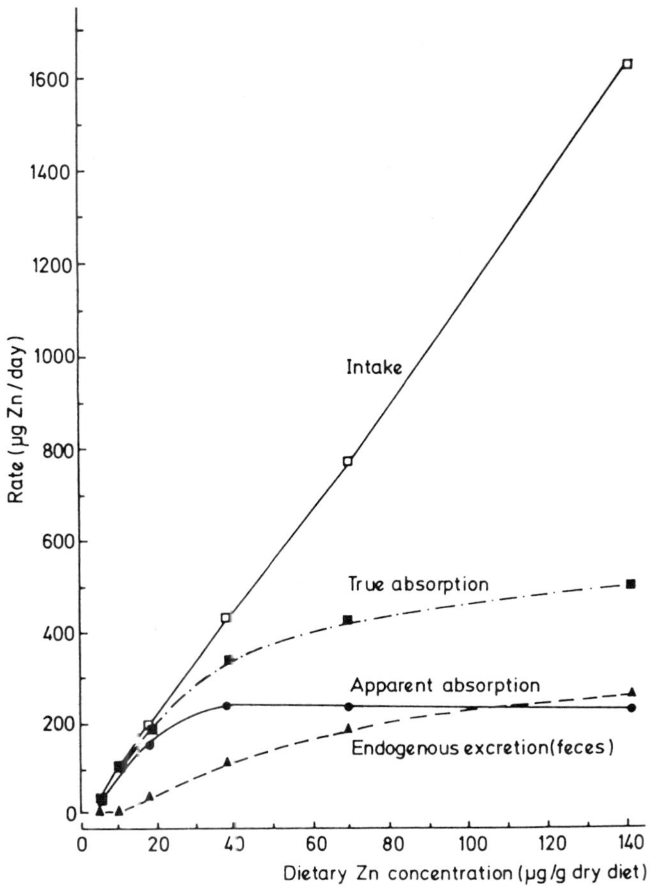

FIGURE 1. True and apparent Zn absorption and fecal excretion of endogenous Zn in relation to dietary Zn supply and intake in young rats. (From Kirchgessner, M. and Weigand, E., *Metal Ions in Biological Systems*, Vol. 15, Siegel, H., Ed., Marcel Dekker, New York, 1983, With permission.)

V. SECRETION OF Zn INTO THE INTESTINE

The importance of endogenous fecal Zn in the homeostatic control of the level of Zn in an animal has been demonstrated by Kirchgessner and Weigand.[2] The intake of Zn was proportional to the concentration of Zn in the diet (Figure 1). However, the curve for true absorption reflected the effect of the controls of absorption already mentioned, in reducing the uptake of Zn by the animal. The endogenous excretion also increased as the concentration of Zn in the diet increased and when this value was subtracted from the true absorption value, the apparent absorption or net retention value resulted. The growth rate of the rats was 0.53, 4.72, 5.58, 6.13, 6.22, and 6.20 g/day for the groups receiving 5.6, 10.6, 18.2, 39, 70, and 141 mg of Zn per kilogram of diet, respectively, indicating that the Zn requirement was approximately 40 mg/kg of diet. It is interesting that the endogenous excretion increased with a small addition of Zn to the diet, considerably prior to meeting the dietary requirement for Zn. It continued to increase after the requirement was met to maintain the apparent Zn absorption constant. There is evidence that the secretion of endogenous Zn into the intestinal tract is an active process.[6] Davies[7] has shown that Zn secretion into ligated loops of rat

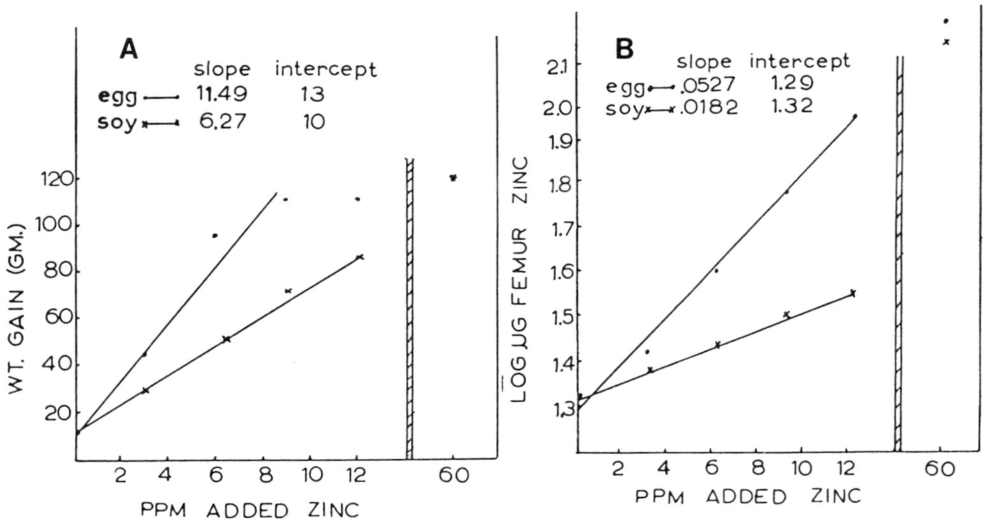

FIGURE 2. Regression of weight gain (A) and log femur (B) on ppm Zn added to a basal egg-white protein diet in the form of Zn carbonate (●) or added as whole fat soy flour (x) substituted for an isonitrogenous amount of egg-white protein. (From Forbes, R. M. and Parker, H. M., *Nutr. Rep. Int.*, 15, 681, 1977. With permission.)

duodenum was not due to secretions from the bile duct, and data from Cousins[1] suggest that metallothionein, which increased in the mucosal cytosol of Zn-treated rats, might be a carrier for the excreted endogenous Zn.

VI. INFLUENCE OF VARIOUS MATERIALS ON Zn AVAILABILITY

A. Soybeans

O'Dell and Savage[8] showed that Zn is essential for chicks when the diet contained isolated soybean protein, but no response to Zn was obtained when the diet contained casein. Morrison and Sarett[9] also observed that the chick requirement for Zn was greater with diets containing isolated soybean protein than those containing casein and gelatin, while Moeller and Scott[10] found that the requirement was lower with an egg white diet than with isolated soybean protein. Norris et al.[11] concluded from their studies that the Zn from soybean protein is relatively unavailable. The improvement in the growth of poults when the isolated soybean protein in the diet was autoclaved[12] was interpreted to indicate that the protein may have bound some of the dietary Zn and that the binding activity was reduced by the heat treatment. The difference in the bioavailability of Zn between whole-fat soy-flour and egg white in the diet of rats is shown in Figure 2 from studies by Forbes and Parker.[13] More supplementary Zn was required to obtain the same growth or femur Zn response with soybean protein in the diet than with egg white.

The bioavailability of Zn in a number of feed and foodstuffs, as compiled by O'Dell,[14] is given in Table 1. The Zn from soy preparations, such as whole-fat soy-flour, seems to be less available than from commercially processed soybean meal.

Approximately 70% of the P in soybeans is in the form of phytic acid. This is associated with the protein in soybean products such as isolated protein. It was shown by O'Dell and Savage[18] that the addition of phytic acid to a purified diet decreased the availability of Zn for chickens. This was also shown to be true for pigs, rats, and man.[19-22] Ca also is involved in the formation of an insoluble complex with phytic acid, protein, and Zn, as is discussed in Chapter 4. Aside from phytic acid, it is probable that protein itself is a contributor to the

Table 1
BIOAVAILABILITY OF Zn IN FEED AND FOODSTUFFS

Product	Assay animal	Method	Value (%)
Milk-based infant formula	Rat	Slope ratio-femur	86
Soy-based infant formula	Rat	Slope ratio-femur	67
Whole-fat soy-flour	Rat	Slope ratio-femur	34
Whole-fat soy-flour	Rat	Slope ratio-femur	54
Soybean meal	Chick	Growth rate	67
Sesame meal	Chick	Growth rate	59
Corn	Chick-rat	Growth rate	63,57
Wheat	Chick-rat	Growth rate	59,38
Rice	Chick-rat	Growth rate	62,39
Egg yolk	Chick-rat	Growth rate	79,76
Nonfat milk	Chick-rat	Growth rate	82,79

problem of the poor availability of Zn. The procedures used in the preparation or processing of soybean products may be responsible for some of the variability in the availability of Zn from these materials.[23] This was investigated extensively by the use of both intrinsically and extrinsically ^{65}Zn-labeled soy-flour, acid-precipitated soy concentrate, and neutralized soy concentrate by Ketelsen et al.[24] The acid-precipitated concentrate was prepared from the soy-flour extract by precipitating at pH 4.5 and freeze-drying while in the neutralized product, the precipitate was brought to pH 8 with the addition of NaOH before freeze-drying. The whole body retention of ^{65}Zn at 1 and 12 days and the retention of ^{65}Zn in the tibia at 12 days indicated that Zn was less available to rats from the neutralized soybean concentrate than from the acid-precipitated soy concentrate or from the soy-flour from which the samples were prepared. There was no difference in the percentage of Zn retained between the intrinsic- and extrinsic-labeled samples containing soy-flour or neutralized soy concentrate (pH 8). There was, however, slightly better use of Zn from the acid-precipitated sample when it was extrinsically labeled with ^{65}Zn, indicating that with that treatment there was not complete mixing of the Zn in the total sample. The conditions of treatment of the soybean product and the presence of other ligands may have a marked effect on the availability of Zn from soybean preparations.

B. Phytate

Phytate is widely distributed in many seeds which are used for food and feed purposes (see Chapter 4). Some seeds, such as rapeseed and sesame seed, may have levels in the 3 to 5% range and some isolated protein fractions may be even higher in phytic acid. As indicated earlier, O'Dell and Savage[18] showed that the addition of phytic acid to a purified diet increases the requirement for Zn and much work since then has confirmed this observation for a number of species. The formation of insoluble complexes of phytic acid, protein, Ca, and Zn has been described in Chapter 4. The absorbability of Zn from a Zn-phytate complex is as high as from equivalent supplements such as the chloride, fumarate, or histidine.[2] The presence of Ca in the diet and the ratio of the various constituents among themselves and to Zn is important. The high content of phytic acid is probably an important factor responsible for the poor availability of Zn from sesame seed meal.[25]

Whether the binding of Zn is due to the phytate or fiber in the food material has been questioned, as it has for Fe. Reinhold et al.[22] showed that in humans the addition of sodium phytate caused increased losses of Zn in the feces and phytate in the form of tanok, an unleavened bread, caused negative Zn and Ca balances. In further studies,[26] they found that the destruction of phytate in the gut was not sufficient to overcome the effects of high phytate intakes.

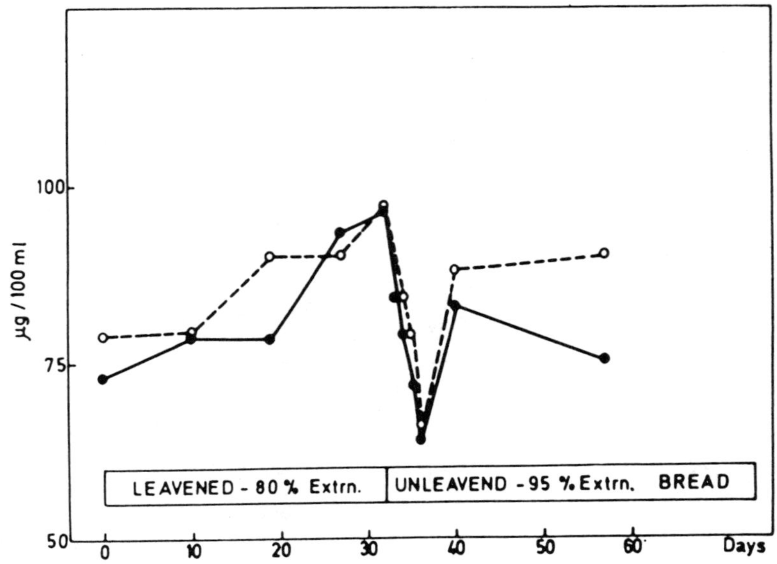

FIGURE 3. The decrease in plasma Zn concentration of two young men following a change from leavened bread made with flour of 80% extraction (bazari) to an unleavened bread made from whole-meals of nearly total extraction is shown. Bread provided 50% of the energy intake of a nutritious hospital diet. The lower limit of plasma zinc concentrations found in well-nourished persons is 75μg/100 mℓ. (From Reinhold, J. G. et al., *Trace Elements in Human Health and Disease*, Vol. 1, Academic Press, New York, 1976. With permission.)

Davies et al.[27] showed that rats consuming bran or phytate exhibited reduced growth and food intake, which was reversed when Zn was supplied in the drinking water. Extracted bran fiber or the addition of a dietary fiber such as pectin had no effect on growth rate and food intake, suggesting that phytate rather than fiber is the major component influencing Zn availability.

The mechanism of phytate interaction with Zn is still an active field for investigation. Forbes et al.[28] found that weight gain and bone ash in rats was not influenced by the Ca level in the absence of phytate, but was progressively depressed by the addition of Ca in the presence of phytate. Graf and Eaton[29] noted a pronounced effect of Ca upon the solubility of Zn complexes with phytic acid.

C. Cereals

With the recognition that cereals contain phytic acid and that the addition of phytic acid to a purified diet can interfere with the utilization of Zn, the assumption was made that phytic acid was responsible for poor utilization of Zn from various cereal-based diets. Reinhold et al.,[30] however, found that the substitution of unleavened, 95% extraction bread for leavened, 80% extraction bread caused a marked drop in plasma Zn concentration in two young male subjects (Figure 3). Since there was a marked difference in the fiber content of these two wheat products of different percentage extraction, they suspected that fiber might play a role in influencing Zn availability, as well as phytic acid. The distribution of ^{65}Zn in an in vitro study showed that white bread, wheat starch, and gliadin bound the most of the Zn in an insoluble form and that preparations of wheat fiber, vegetable fiber, and cellulose held somewhat less, but still considerable amounts of Zn in insoluble form. The Zn is released from the white bread, wheat starch, and gliadin when these materials are digested, but less so from the indigestible fiber, which remains intact in the GI tract (Figure 4). When 10 g of cellulose were consumed daily for 14 days by two young men, there was

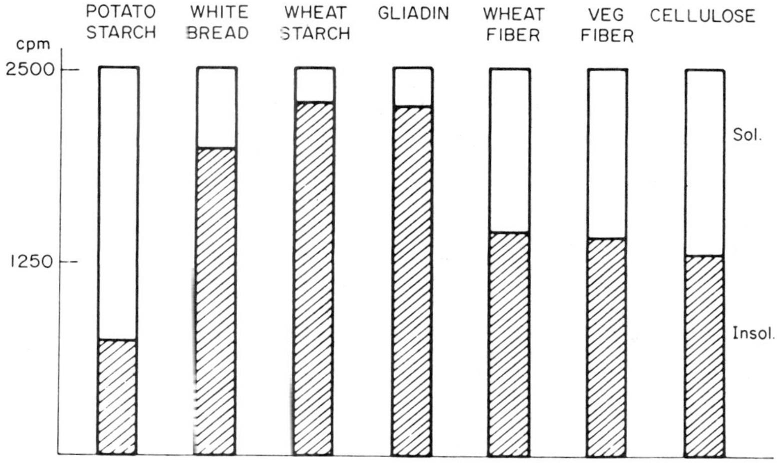

FIGURE 4. The distribution of ^{65}Zn when 50 mg of each of the solids included was suspended in a saline-glucose-glycine solution containing approximately 2500 cpm of ^{65}Zn per 3 mℓ and a total Zn concentration of 75 μg/100 mℓ. The solution also contained 8.5 g NaCl, 0.3 g KCl, 3.6 g glucose, 0.375 g glycine, and 1.3 m*M* phosphate per liter at pH 6.8. After rotation on a tube rotator for 90 min, the tubes were centrifuged and 3 mℓ of supernate taken for counting in an automatic gamma counter. (From Reinhold, J. G. et al., *Trace Elements in Human Health and Disease*, Vol. 1, Prasad, A. S., Ed., Academic Press, New York, 1976. With permission.)

an increase in the loss of Zn in the feces, which led to a negative Zn balance. Plasma Zn and serum P levels were also reduced. This study suggested that fiber as well as phytic acid might be important in influencing Zn availability in cereals.

Absorption of Zn in man was studied with diets containing bread with 100 or 72% extraction.[31] A lower absolute amount of Zn was absorbed from the white bread compared to the whole wheat bread. Zn from added Zn chloride was more available from the white bread than from the whole wheat bread. The addition of milk, cheese, beef, and egg improved the absorption of Zn from the diet containing the whole wheat bread.

Reinhold et al.[32] conducted balance studies on two men consuming either a high-fiber diet based on whole-meal wheat bread or white bread for 20-day periods. The subjects consuming whole wheat bread were in negative balances for Ca, Mg, Zn, and P due to increased fecal excretion of each element. The fecal losses correlated with the fecal dry matter which was influenced, in turn, by the fecal fiber excretion. These elements were well utilized by subjects who consumed white bread as they were in positive balance. In still another study,[33] these workers added 10 g of cellulose dispersed in apple compote to the daily diet of three men for a 20-day period. The losses of Ca and Zn were greater than their consumption and their levels in the plasma were also reduced. Although some variation was found between subjects, the authors concluded that the results indicated that high intakes of fiber could explain, to a great extent, the poor utilization of Ca, Zn, and Mg by the villagers in Iran where high-fiber diets are consumed.

The importance of fiber in binding Zn is still controversial. Some studies suggest little or no effect of fiber from various sources on Zn availability. Guthrie and Robinson[34] found no overall change in Zn metabolism following supplementing the diet of women with wheat bran, although Zn balance and serum Zn levels declined in some subjects. In other studies, the addition wheat bran, corn bran, and soybean hulls to bread had no effect on Zn metabolism,[35] but the fiber from fruits and vegetables produced a negative Zn balance.[36] Cellulose added to a casein-based rat diet caused no interaction with tissue Zn in the liver, tibia, or

Table 2
FEMUR Zn CONCENTRATIONS OF RATS FROM SELECTED DIET GROUPS (EXPERIMENT 5)

Phytate to Zn molar ratio	Femur Zn concentration[1] (ppm)
0[2]	190 ± 7.8[1,4]
14	212 ± 11[1]
25	114 ± 4.2[2]
34	85 ± 2.7[3]
20 + Ca[3]	108 ± 12[3]
25 + Ca	78 ± 3.0[3]

[1] Dry weight basis.
[2] 12 ppm Zn as $ZnSO_4 \cdot 7H_2O$ and no phytate. All other diets contained soft wheat bran to supply 12 ppm Zn and molar ratio of 52. Lower molar ratios were attained by addition of $ZnSO_4 \cdot 7H_2O$.
[3] 1% additional Ca as $CaHPO_4$; all other diets contained 0.75% Ca supplied by basal ingredients.
[4] Mean ± SE for six animals; means that share common superscripts are not significantly different ($p < 0.05$).

From Morris, E. R. and Ellis, R., *J. Nutr.*, 110, 2000, 1980. With permission.

testes.[37] The 10% addition of nondigestible polysaccharides such as carrageenan, agar-agar, and guar gum reduced Zn absorption while sodium alginate had no effect.[38]

It is very difficult to separate the effects of fiber and phytic acid because these constituents, in general, have a similar distribution pattern in foods. There also may be adaptation to the feeding of fiber so that some of the studies were of too short a duration to reach equilibrium. It may as well be assumed that both phytic acid and fiber are responsible for the poor availability of Zn in foods.

Morris and Ellis[39] were able to reduce the phytate content of wheat bran by extraction or enzymatic treatment. The availability of Zn as measured by growth or femur Zn content of rats was greatly improved by the reduction in phytate level. The molar ratio of phytate to Zn has been used to indicate that extent to which Zn may be complexed. The level of Ca, as indicated in Table 2, can greatly influence the availability of the Zn in the presence of phytate. The ratio also does not account for the fact that phytate in wheat bran does not affect Zn utilization to the same extent as sodium phytate added to a diet, and similar phytate to Zn ratios may have different Zn availabilities, depending upon the source of the phytic acid. Lantzsch et al.[40] were able to increase the Zn content of wheat and barley by fertilization of the soil with Zn (Table 3). The availability of the Zn for rats was significantly improved with the higher level of Zn in the wheat and barley, which reduced the phytate to Zn ratio.

In a study of Zn absorption from composite meals of conventional foods,[41] Zn absorption from protein-rich meals seemed to depend more upon the total amount of Zn in the meal than on the presence of phytic acid or Ca.

Table 3
RELATIVE AVAILABILITY OF Zn FOR ABSORPTION ($ZnSO_4$ = 100) FROM WHOLE WHEAT AND WHOLE BARLEY AFTER Zn FERTILIZATION OR SUPPLEMENTATION (6 ANIMALS PER GROUP)

	Wheat		Barley	
Treatments	**Molar ratio**	**Zn availability (%)**	**Molar ratio**	**Zn availability (%)**
Unfertilized	43	59[a]	34	72[a]
Fertilized	25	75[b]	28	76[a]
Supplemented		87[c]		78[a]
Fertilized	19	82[b,c]	22	81[a]
Supplemented		83[b,c]		88[b]

Note: Values not sharing a common superscript are significantly different ($p < 0.05$).

From Lantzsch, H. J. et al., *Trace Element Metabolism in Man and Animals,* Proc. 4th Int. Symp. (TEMA 4), Gawthorne, J. M., McC. Howell, J., and White, C. L., Eds., Springer-Verlag, Basel, 1982, 107. With permission.

D. Legumes and Oil Seed Meals

The availability of Zn in soybean protein has been studied extensively and to a lesser extent in other legumes and oil seeds. Phytic acid is present in these materials and has been the main focus of these studies, as reviewed by Maga.[42]

Allred et al.,[43] in studying the availability of Zn from isolated soybean protein, developed an in vitro test for Zn-binding activity. The isolated soybean protein sample was suspended in a solution of ^{65}Zn, shaken, centrifuged, and the radioactivity of the supernatent solution determined in a gamma counter. The binding activity increased with increasing pH and the addition of sodium phytate increased the binding activity. A sample of isolated soybean protein in which the major portion of the phytic acid was removed showed reduced Zn-binding activity. Binding activity was reduced when isolated soybean protein was autoclaved. Alpha casein, as well as isolated soybean protein, had Zn-binding activity.

Lease and Williams[44] used a modification of this method to study the Zn-binding activity of safflower meal, soybean meal, isolated soybean protein, cottonseed meal, and several samples of sesame seed meal. The samples were also tested for Zn availability by chick feeding studies. The in vitro test gave qualitative results similar to the in vivo results. A sample of sesame meal which was particularly growth depressing on a low-Zn diet was improved by autoclaving and the in vitro Zn-binding was reduced. Interestingly, a relationship between phytate content of the meals and availability of Zn by both the in vitro and in vivo methods was not found for these samples.

Rapeseed has been improved nutritionally though an intensive breeding program in Canada. A protein concentrate from rapeseed (Tower variety of *Brassica napus*) was observed to give improved growth of rats when they were given Zn in their drinking water.[45] The protein concentrate contained about 5.7% phytic acid and 120 mg/kg of Zn (molar ratio of 47). In further studies with rats, it was necessary to supplement with approximately 150 to 200 mg Zn per kilogram of diet to maintain normal serum and femur Zn levels. This was in excess of the amount of Zn needed to bring the phytic acid to Zn ratio to 30, a figure usually regarded as a safe ratio. This indicates, again, that several factors may alter the phytic acid and Zn relationship.

E. Coffee and Dairy Products

Several foods were tested by Pecoud et al.[46] for their effect of the availability of Zn from zinc sulfate. The Zn salt was fed with a meal of the test material to human subjects and the serum Zn concentration was measured up to 9 hr afterwards. Zn absorption was decreased with milk and cheese, brown bread, and also by purified phosphate or phytate in the same amounts as found in the foods. Of interest, also was the fact that coffee, without milk or sugar, seemed to inhibit Zn absorption. The serum Zn level after administering coffee with 50 mg of Zn was equal to that with 25 mg of Zn alone.

F. Oysters

Oysters are among the foods richest in Zn; about half of it is dialyzable and loosely bound while the other half is tightly bound.[3] One might expect that the large amount of bound Zn would influence the biological availability of the total Zn in this food. In nutritional studies, however, Zn from lyophilized oyster powder and Zn carbonate were equally available.[47] Solomons,[3] however, found that an excess of Fe did not interfere with Zn availability from oysters as it does with Zn from inorganic sources.

G. Synthetic Chelates

In a study of the effect of autoclaving of isolated soybean protein on the availability of its Zn, as well as the availability of Zn from the synthetic chelating agent, EDTA, was tested.[12] As shown in Table 4, not only was the Zn available from the Zn-EDTA complex, but the addition of EDTA also improved the availability of Zn from the isolated soybean protein as measured by the growth of poults fed the basal diet without any supplementary Zn. There was no response to EDTA when Zn was adequately supplied and the response was not due to a reduction in the available Ca since lowering the Ca in the diet had no effect upon poult growth. The mechanism for the action of EDTA was not apparent, but it was proposed that the EDTA was removing Zn from its complex with isolated soybean protein and that Zn-EDTA was a small enough molecule to be absorbed from the GI tract.

In a study of the quantitative relationship of EDTA to dietary Zn in poults,[48] various levels of EDTA and Zn were added to a poult diet containing isolated soybean protein. In supplementing the basal diet, 100 mg of EDTA per kilogram were equivalent to approximately 8 mg/kg of Zn. When 7 or 10 mg/kg of Zn was added to the diet, 100 mg of EDTA per kilogram was equivalent to approximately 19 mg of Zn per kilogram. The data indicated that the chelating agent was not only improving the availability of the Zn present in the isolated soybean protein, but was also improving the utilization of the suboptimal amount of Zn which was added to the basal diet.

The mechanism by which EDTA influences the absorption of Zn was studied in isolated intestinal segments in chickens.[47] Complexes of radioactive Zn or Ca with ^{14}C-EDTA were placed in intestinal segments and mesenteric venous blood was collected at 20-min intervals for 140 or 180 min. The parallel appearance of ^{14}C with ^{65}Zn and ^{45}Ca suggested that EDTA and the cations pass through the intestinal mucosa together (Figure 5). This was in contrast to earlier reports that very little of the ^{14}C activity of an orally administered EDTA-^{14}C appeared in the blood of rats and man.[50] Vohra and Gonzales[51] studied the specificity of EDTA on aiding cation absorption. Mixtures of radioactive Zn and Mn were introduced into a segment of the GI tract of poults and the loop returned to the body cavity for a 2-hr period. The ratio of Zn to Mn was determined both in the residual material in the isolated segment and in the blood. When EDTA was added to the Zn-deficient or Zn-supplemented diets, it

Table 4
THE EFFECT OF EDTA ON THE GROWTH AND INCIDENCE OF PEROSIS IN POULTS FED RATIONS CONTAINING ISOLATED SOYBEAN PROTEIN[a] AND VARIOUS LEVELS OF Zn

Added Zn (ppm)	Added EDTA[b] (ppm)	Body weight gain and perosis score[c] Exp. 6 (g)	Exp. 7 (g)	Exp. 8 (g)	Exp. 9 (g)
0	0			170(2.6)	162(2.4)
11	0	218(2.9)	184(2.9)		
20					259(2.9)
30	0	261(2.4)	308(3.1)		
40					272(1.8)
60	0			315(1.0)	
150	0				288(0.8)
0	227			238(3.0)	308(0.9)
11	227	300(1.8)	311(3.0)		
11	454		320(1.3)		
30	227[d]	287(0.9)	339(1.2)		
20	227				295(1.4)
40	227				294(0.8)
60	227			327(1.0)	
150	227				297(1.2)
11	0		219(2.9)[e]		
Duration	(days)	18	19	19	19

[a] Drackett C-1.
[b] Ethylenediaminetetracetic acid, disodium salt.
[c] Perosis score within parentheses.
[d] Fed as Zn complex equivalent to 19 ppm Zn.
[e] $CaCO_2$ in diet reduced by 1%.

From Kratzer, F. H. et al., *J. Nutr.*, 68, 313, 1959. With permission.

enhanced the absorption of Zn from the intestinal segment to the blood, as shown by lower isotopic ratios in the segments and higher ratios in the blood. This again demonstrates that effect of EDTA in improving the absorption of Zn and shows that the effect upon Zn is greater than upon Mn.

Forbes[52] observed the effect of EDTA in improving the utilization of Zn in rats fed an isolated soybean diet. EDTA increased weight gain, Zn absorption and retention, and concentration of Zn in the femur. Oberleas et al.,[53] also working with rats, observed that EDTA improved the absorption of Zn. The absorption was impaired by the addition of phytic acid and Ca to the diet because of the formation of a poorly soluble complex of Zn, phytic acid, and Ca. It was uncertain whether protein was also involved in this insoluble complex. Likuski and Forbes[54] observed that the addition of phytic acid to either an amino acid or a casein diet marginal in Zn reduced the growth of chickens. With both the amino acid and casein diets, growth was greatly improved by the addition of EDTA.

A whole body counting method was used by Suso and Edwards[55] to study Zn availability and the biological half-life in chickens. The method involved administering ^{65}Zn either orally or intraperitoneally and determining the isotope in the bird initially and at 2-day intervals up to 10 days (Figure 6). The amount of isotope unabsorbed is excreted by 2 days so that the level in the body found at 2 days and after may be extrapolated to zero time to give a

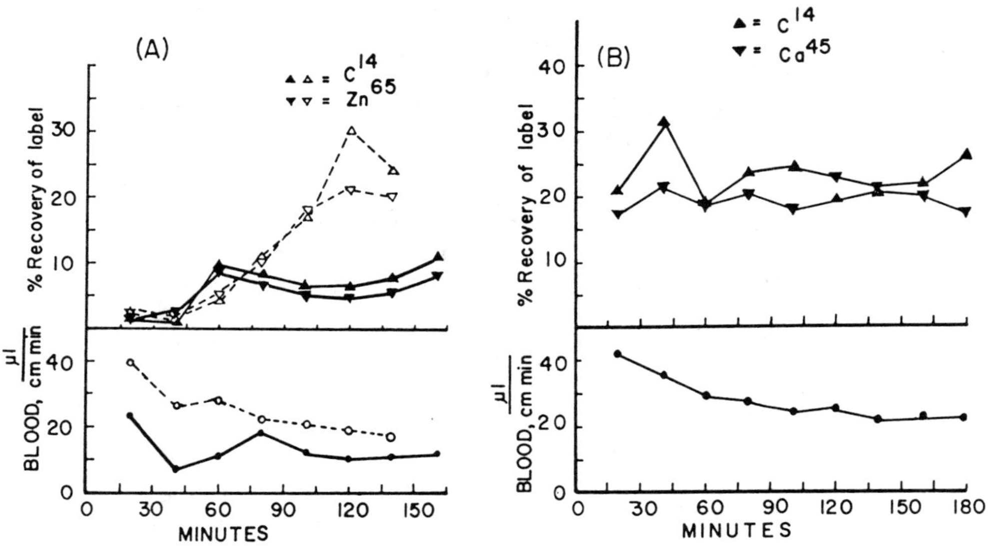

FIGURE 5. Blood flow and rate of appearance of ^{14}C and ^{65}Zn or ^{45}Ca in the blood perfusing isolated intestinal segments. Rate of appearance expressed as percent of injected label appearing in blood per cm $min^{-1} \times 10^3$. (From Koike, T. L. et al., *Proc. Soc. Exp. Biol. Med.*, 117, 483, 1964. With permission.)

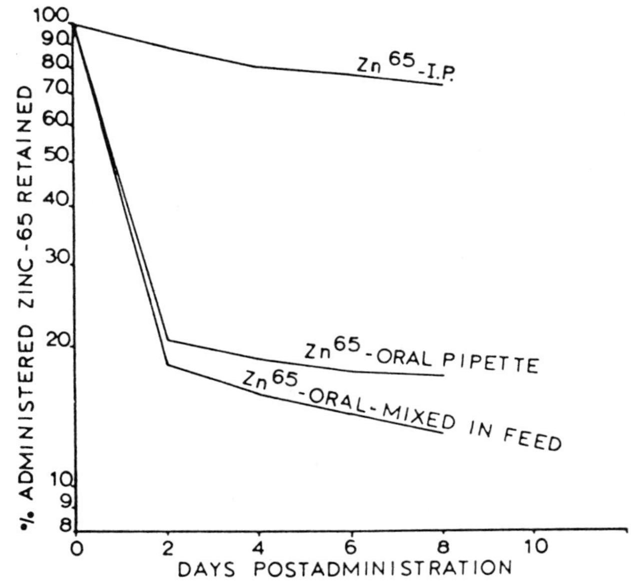

FIGURE 6. Retention of ^{65}Zn after oral administration of the isotope by pipette into the crop or by mixing in the feed. (From Suso, F. A. and Edwards, H. M., Jr., *Poultry Sci.*, 47, 1417, 1968. With permission.)

measure of the amount of Zn absorbed. The slope of the curve for retained Zn is a measure of the biological half-life. ^{65}Zn absorption was significantly increased as levels of EDTA were increased from 0 to 0.165% of the diet, with the maximum absorption observed at 0.125%. The biological half-life of Zn was reduced only from 18.9 to 15.3 days by 0.165% EDTA while that of Fe, for example, was reduced from 27.4 to 13.6 days by 0.24% EDTA.[56]

The effectiveness of several chelating agents in improving the availability of Zn was measured with the turkey poult.[57] The basal diet contained isolated soybean protein and was deficient in Zn. The chelates were fed at a level of 0.684 mmol/kg of diet for about 20 days. The improvement in the growth of the poults in comparision to the basal diet alone was used as a measure of the effectiveness of the chelating agent to improve the availability of Zn from the basal diet. The relative improvement in growth for the 28 chelating agents shown in Table 5 was determined in several different feeding tests.

There was considerable variation in the activity of the various chelating agents in improving Zn availability. Compounds such as aminonaphtholsulfonic acid (ANS), diaminocyclohexanetetraacetic acid (CDTA), dihydroxyethylglycine (DHEG), dihydroxybenzenedisulfonic acid (DHBDS), ethylenediaminetetrapropionic acid (EDTP), ethylenediaminetetraethanol (EDTE), glutamic acid (GA), ethylenediaminebitartrate (EDBT), iminodiacetic acid (IDA), hydroxyquinolinesulfonic acid (HQS), mercaptosuccinic acid (MSA), proline (P), thalicomide (TA), triethylenetetraamine (TETA), terephthalic acid (TP), and thiodipropionic acid (TPA) gave growth responses which were either poorer than on the Zn-deficient basal diet or of the same order.

On the other hand, compounds such as nitrilotriacetic acid (NTA), hydroxyethylethylenediaminetriacetic acid (HEDTA), propylenediaminetetraacetic acid (PDTA), hydroxypropylenediaminetetraacetic acid (HPDTA), dihydroxyethylethylenediaminediacetic acid (DHEEDA), ethylenediaminediacetic acid dipropionic acid (EDDADP), and ethylenediaminediacetic acid (EDDA) gave definite improvement in the growth of poults and the relative growth for some of the chelating agents was of the same order as for EDTA.

Improved growth, but definitely poorer than that for EDTA, was observed for ethylenebis (oxyethylenenitriolo) tetraacetic acid (EBONTA), hydroxyethyliminodiacetic acid (HEIDA), and diethylenetriaminepentaacetic acid (DTPA). High mortality was obtained with ethylenediamine (dihydroxyphenyl) acetic acid (EDDHA) and it was difficult to evaluate the response.

The growth response of some of the compounds is compared with their stability constants as taken from the literature in Chapter 2, Figure 14. The various compounds plotted varied from stability constants of 5.36 for DHEG to 18.67 for CDTA. One should bear in mind, however, that the stability constants used are maximum values which may be obtained under highly basic conditions (which would never exist physiologically).

The growth response appeared to be equally good over the range of stability constants from 13 to 17, with an optimum value at 14.5. The greatest responses were obtained with EDTA, EDDADP, and HEDTA, which have stability constants of 16.5, 14.5, and 14.5, respectively. NTA, with a stability constant of 10.45, gave a definite growth response. The reduction of one or more of its carboxyl groups (HEIDA, DHEG) reduced the stability constants and also the biological response. EDDA improved the availability of Zn more than NTA. A modification of EDDA to EDTA by the addition of two more acetic acid molecules increased the stability constant for Zn from 11.1 to 16.5. The reduction of one of the acetic acids for EDTA results in HEDTA, which has a stability constant of 14.5. Attachment of two molecules of propionic acid to EDDA gives rise to EDDADP, with a stability constant of 14.5. These two compounds, both with stability constants of 14.5, were the most effective in improving Zn availability in this study. Acetic acid molecules attached to ethylenediamine tend to give chelating agents with stability constants higher than propionic acid. EDTP has a value of 7.8 and has poor biological activity.

When all the acetic acid molecules of EDTA were reduced to ethanol, EDTE was obtained, which did not improve the availability of Zn. However, the compound containing two alcohol and two acetic acid molecules attached to ethylenediamine (DHEEDA) gave a relative growth response of the same order as EDTA. No data are available about its stability constant.

Compounds like EBONTA and TETA were less effective for improving Zn availability

Table 5
THE CHELATING AGENTS: THEIR STRUCTURES, ABBREVIATED NAMES, MOLECULAR WEIGHTS, AND STABILITY CONSTANTS FOR Zn

No.	Chelating agent	Structure	Mol wt	Abbreviation	Stability constant for Zn
1	1-Amino-2-naphthol-4-sulfonic acid[a]	$NH_2C_{10}H_5(OH)SO_3H$	239	ANS	
2	1,2-Diaminocyclohexanetetraacetic acid[b]	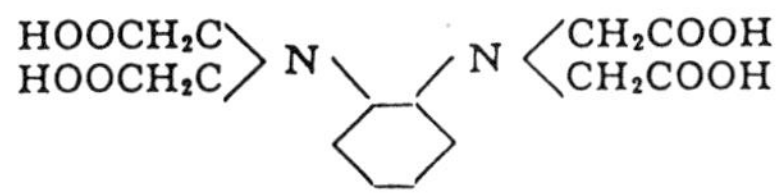	340	CDTA	18.67
3	Dihydroxyethylethylenediaminediacetic acid[c]	$(HOH_2CH_2C)(HOOCH_2C)N—CH_2—CH_2—N(CH_2CH_2OH)(CH_2COOH)$	264	DHEEDA	
4	Diethylenetriaminepentaacetic acid[d]	$(HOOCCH_2)_2NCH_2CH_2—N(CH_2COOH)—CH_2CH_2N(CH_2COOH)_2$	393	DTPA	18.14
5	4,5-Dihydroxy-*m*-benzenedisulfonic acid disodium salt[a]	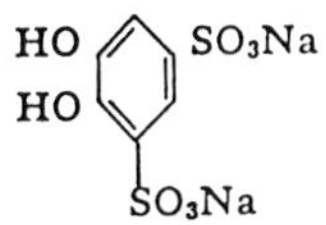	314	DHBDS	10.41
6	*N-N*-Di(β-hydroxyethyl) glycine (Na salt)[d]	$(HOH_2CH_2C)_2NCH_2COONa$	185	DHEG	5.36
7	[Ethylenebis(oxyethylenenitrilo)] tetraacetic acid[a]	$(HOOCH_2C)_2—N—CH_2CH_2—O—CH_2—CH_2—O—CH_2CH_2N(CH_2COOH)_2$	380	EBONTA	11.0

8	Ethylenediaminebitartrate[b]	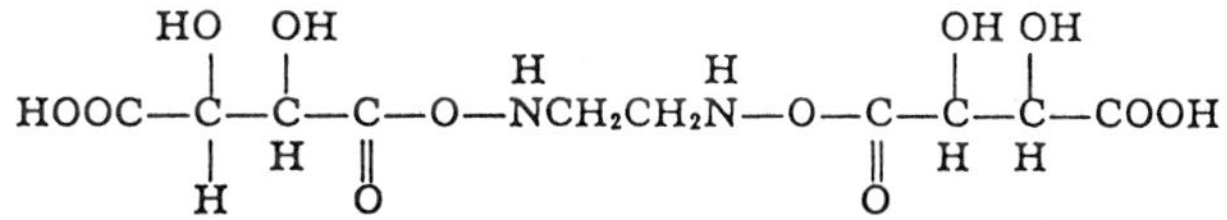	324	EDBT	
9	Ethylenediamine-*N,N′*-diacetic acid[b]	$HOOCH_2C{-}NH{-}CH_2CH_2{-}NHCH_2COOH$	176	EDDA	11.1
10	Ethylenediamine-*N,N′*-diacetic acid-*N,N′*-dipropionic acid[e]	$(HOOCH_2C)(HOOCH_2CH_2C)N{-}CH_2CH_2{-}N(CH_2COOH)(CH_2CH_2COOH)$	318	EDDADP	14.5
11	Ethylenediamine (di(*O*-hydroxyphenylacetic acid))	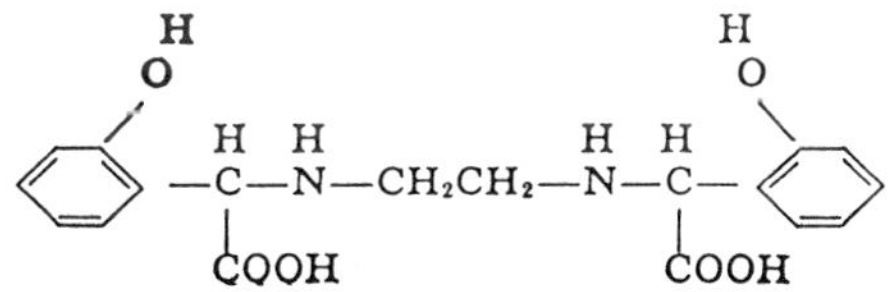	361	EDDHA	9.26
12	Ethylenediaminetetraacetic acid[a]	$(HOOCH_2C)_2NCH_2CH_2N(CH_2COOH)_2$	292	EDTA	16.5
13	Ethylenedinitrilotetraethanol[a]	$(HOH_2CH_2C)_2NCH_2CH_2N(CH_2CH_2OH)_2$	236	EDTE	
14	Ethylenediaminetetrapropionic acid (tetrasodium salt)	$(NaOOCH_2CH_2C)_2NCH_2CH_2N(CH_2CH_2COONa)_2$	436	EDTP	7.8
15	Glutamic acid	$HOOCH_2CH_2C{-}CH(NH_2){-}COOH$	147	GA	5.45
16	(2-Hydroxyethylimino)diacetic acid[a]	$HOH_2CH_2C{-}N(CH_2COOH)_2$	177	HEIDA	8.57
17	Hydroxyethylethylenediaminetriacetic acid[a]	$(HOH_2CH_2C)(HOOCH_2C)N{-}CH_2CH_2N(CH_2COOH)_2$	278	HEDTA	14.5

Table 5 (continued)
THE CHELATING AGENTS: THEIR STRUCTURES, ABBREVIATED NAMES, MOLECULAR WEIGHTS, AND STABILITY CONSTANTS FOR Zn

No.	Chelating agent	Structure	Mol wt	Abbreviation	Stability constant for Zn
18	2-Hydroxypropylenediaminetetraacetic acid[f]	$(HOOCH_2C)_2N—CH_2CH(OH)CH_2N(CH_2COOH)_2$	322	HPDTA	
19	8-Hydroxy-5-quinoline-sulfonic acid[a]	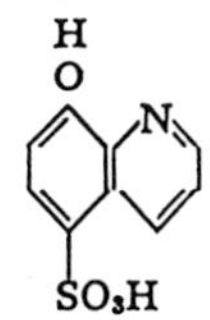	225	HQS	8.72
20	Iminodiacetic acid disodium salt monohydrate[a]	$HN(CH_2COONa)_2 \cdot H_2O$	195	IDA	7.02
21	Mercaptosuccinic acid[a]	$HOOC—CH_2—CH(SH)—COOH$	150	MSA	
22	Nitrilotriacetic acid disodium salt	$HOOCH_2C—N(CH_2COONa)_2$	235	NTA	10.45
23	Propylenediaminetetraacetic acid tetra Na salt	$(NaOOCCH_2)_2NCH_2CH_2CH_2N(CH_2COONa)_2$	394	PDTA	
24	Proline	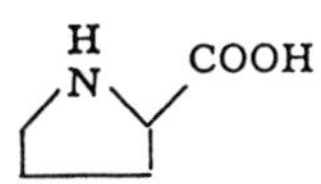	115	P	10.2

25	Terephthalic acid	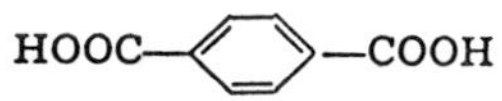	166	TP	
26	Thalidomide	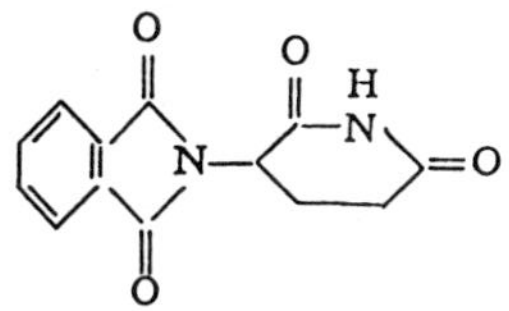	258	TA	
27	Thiodipropionic acid[g]	$S(CH_2CH_2COOH)_2$	178	TPA	
28	Triethylenetetraamine	$H_2NCH_2CH_2NH—CH_2CH_2—NH—CH_2CH_2NH_2$	146	TETA	12.1

[a] Eastman Organic Chemicals, Rochester, N.Y.
[b] K and K Laboratory, Inc., Jamaica 33, N.Y.
[c] Courtesy of Dr. A. Wallace, UCLA, Los Angeles, Calif.
[d] Courtesy of Dow Chemical Company, Midland, Mich.
[e] Laboratory synthesis carried out in the investigator's laboratory.
[f] Aldrich Chemical Company, Inc., Milwaukee 10, Wisc.
[g] Courtesy of American Cyanamid Company, N.Y.
[h] Courtesy of Geigy Chemical Corporation, Ardsley, N.Y.

than would be predicted on the basis of their stability constants. CDTA and DTPA, with stability constants over 18, were also relatively ineffective in improving the availability of Zn. It was proposed that these compounds bound Zn so tightly that it was unavailable to the animal after being absorbed. It has subsequently been shown that the Zn is available from its CDTA-Zn complex, so the explanation for the low biological activity is not apparent.

The effectiveness of several synthetic chelating agents in improving the availability of Zn for chicks was studied by Nielsen et al.[58] They used a low-Zn basal diet containing isolated soybean protein. The studies were carried out for 4 weeks and data were recorded on body weight, feather score, leg score, length to narrow width ratios of the femurs, and the Zn content of the tibias in one experiment. The results, as shown in Table 6, were in good agreement with the previously mentioned work with poults. EDDA gave a slightly better response than with poults, but the most effective compounds were HEDTA, EDTA, and DTPA. Both cysteine and histidine added at 0.5% of the diet were partially effective. In general, there was good agreement between the responses in growth, feather score, leg score, and Zn content of the tibia. One exception was with histidine, which gave a good response in growth, feather score, and bone disorder, but had little or no effect on the Zn content of the tibia. Tryptophan, anthranilic acid, xanthurenic acid, kynurenic acid, cystine, and glutamic acid were essentially ineffective in influencing Zn utilization in these experiments.

The effect of various synthetic chelating agents was also studied by Suso and Edwards[56] with chicks using their whole body counting system. HEIDA was as effective as EDTA in improving Zn availability at equivalent molar levels, while DTPA was effective at half the molar equivalent. In general, they could not relate the effectiveness of the chelating agent to the stability constants obtained from the literature.

In studies with the isolated, vascularly perfused rat intestine, only EDTA was effective in significantly increasing the transfer of Zn from the intestinal lumen to the portal circulation.[59] The other compounds tested (citric acid, cysteine, reduced glutathione, histidine, methionine, picolinic acid, and tryptophan) were without activity.

Since Zn may be coprecipitated as Zn-Ca-complex, Welch et al.[60] reasoned that Zn might also be coprecipitated with Ca-oxalate in a food material rich in oxalic acid. Rats fed Zn-deficient diets with and without sodium oxalate were orally dosed with either ^{65}Zn-labeled spinach leaves or ^{65}Zn-labeled zinc sulfate. Dietary oxalate enhanced the availability of ^{65}Zn from zinc sulfate but had no effect on absorption and retention of ^{65}Zn from spinach leaves. The absorption of Zn was greater from the spinach leaves than from zinc sulfate and there was no evidence for a deleterious effect of oxalic acid on Zn availability.

Phenytoin, an anticonvulsant drug with two oxygen and nitrogen

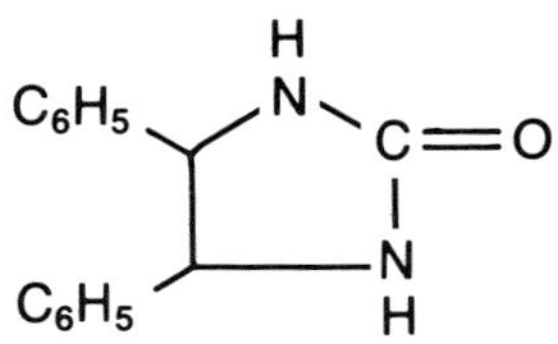

Phenytoin

atoms, readily chelates divalent cations.[61] In studies with rats fed a standard rat diet, absorption of ^{65}Zn was improved by about 40% with the daily administration of 10 to 100 mg of this material. There was also a greater daily loss of ^{65}Zn when the phenytoin was given. The levels of the drug tested in rats were far in excess of the comparable dose administered to humans, so it is not known whether Zn metabolism is altered in patients to whom phenytoin is administered.

Table 6
BODY WEIGHTS, FEATHER SCORES, LEG SCORES, LENGTH-TO-NARROW WIDTH RATIOS (L/NW) OF FEMURS, AND Zn CONTENT OF TIBIAS AT 4 WEEKS OF CHICKS FED A SOY PROTEIN DIET WITH SYNTHETIC AND NATURAL CHELATING AGENTS

Chelating agent	Zn supplemented (ppm)	Body weight (g)	Feather score	Leg score	L/NW femur ratio		Zinc in tibias: Ends (ppm)[a]	Zinc in tibias: Shafts (ppm)[a]
Experiment 1								
None	5[2]	141(0)[3ab4]	4.3[a5]	4.5[a5]	9.1[ab]		33[a]	45[a]
None	40	284(10)[d]	1.2[c]	1.2[c]	11.6[cd]		97[c]	186[d]
DHEG(5.36)[6,7]	5	143(10)[ab]	3.9[a]	4.1[ab]	9.4[ab]		41[a]	60[a]
IDA(7.02)[7]	5	139(0)[ab]	4.3[a]	4.0[ab]	9.1[ab]		35[a]	45[a]
HEIDA(8.57)[7]	5	191(0)[c]	2.8[b]	3.5[b]	9.5[ab]		36[a]	51[a]
EDDHA(9.26)[7]	5	156(70)[abc]	3.0[ab]	2.3[c]	11.2[cd]		42[ab]	66[a]
EBONTA(11.00)[7]	5	167(10)[bc]	4.1[a]	4.3[a]	8.6[a]		37[a]	50[a]
EDDA(11.10)[7]	5	256(30)[d]	1.7[c]	1.4[c]	11.8[d]		55[bc]	97[b]
HEDTA(14.50)[7]	5	291(0)[d]	1.5[c]	1.3[c]	10.8[cd]		70[d]	132[c]
EDTA(16.50)[7]	5	280(30)[d]	1.6[c]	1.3[c]	11.3[cd]		74[d]	131[c]
DTPA(18.14)[7]	5	283(30)[d]	1.6[c]	1.1[c]	11.3[cd]		58[c]	100[b]
CDTA(18.67)[7]	5	102(40)[a]	4.8[a]	4.3[ab]	10.2[bc]		36[a]	54[a]
Experiment 2								
None	5	137(10)[ab]	4.0[a]	4.2[ab]	9.2[abc]	32[a]		
None	80	278(0)[de]	1.5[c]	1.3[e]	11.7[gh]	142[e]		
Try(9.30)[8]	5	141(0)[ab]	4.5[a]	4.9[a]	8.6[a]	32[a]		
Try(9.30)[8]	80	292(0)[ef]	1.5[c]	1.3[e]	11.5[gh]	148[ef]		
AA(20.93)[7]	5	170(0)[b]	4.1[a]	4.3[ab]	8.9[ab]	30[a]		
AA(20.93)[7]	80	318(0)[fg]	1.4[c]	1.1[e]	11.6[gh]	137[e]		
XA[7]	5	137(0)[ab]	4.3[a]	4.2[b]	9.4[bcd]	33[a]		
XA[7]	80	292(0)[ef]	1.3[c]	1.4[e]	11.9[h]	142[e]		
KA[7]	5	149(0)[ab]	4.5[a]	4.6[ab]	8.8[ab]	30[a]		
KA[7]	80	298(0)[efg]	1.8[c]	1.5[de]	11.0[fg]	141[e]		
CySH(18.20)[8]	5	228(0)[c]	1.7[c]	2.3[c]	10.1[de]	45[b]		
CySH(18.20)[8]	80	328(10)[c]	1.6[c]	1.2[e]	11.8[gh]	186[g]		
Cys[8]	5	146(0)[ab]	4.1[a]	4.1[b]	9.3[abc]	31[a]		
Cys[8]	80	305(0)[efg]	1.4[c]	1.2[e]	11.2[gh]	158[f]		
His(12.88)[8]	5	156(0)[ab]	2.9[b]	2.1[cd]	10.4[ef]	35[ab]		
His(12.88)[8]	80	315(0)[fg]	1.4[c]	1.3[e]	11.5[gh]	179[g]		
HEDTA(14.50)[7]	5	259(0)[d]	1.4[c]	1.2[e]	11.7[gh]	95[c]		
HEDTA(14.50)[7]	80	305(0)[efg]	1.6[c]	1.1[e]	11.4[gh]	157[f]		
Glu(9.46)[8]	5	136(0)[a]	4.5[a]	4.2[b]	9.7[cde]	32[a]		
Glu(9.46)[8]	80	325(0)[g]	1.4[c]	1.3[e]	11.4[gh]	125[d]		

1 Fat-free, dry basis.
2 Diet contained a total of 12 ppm Zn by analysis.
3 Mortality rate in percent.
4 For each experiment, values within a column followed by the same letters are not significantly different ($p < 0.05$) from each other.
5 1 = Normal; 5 = very abnormal.
6 Stability constant for Zn, if known, indicated in parenthesis. Meanings of abbreviations: DHEG, *N,N*-di(2-hydroxyethyl)glycine, Na salt; IDA, iminodiacetic acid salt monohydrate; HEIDA, (2-hydroxyethylimino)-diacetic acid; EDDHA, ethylenediamine (di (*o*-hydroxyphenylacetic acid)); EBONTA, (ethylenebis (oxyethylene-nitrilo)tetraacetic acid; EDDA, ethylene-*N,N'*-diacetic acid; HEDTA, hydroxyethylethylenediaminetriacetic acid; EDTA, ethylenediaminetetraacetic acid; DTPA, diethylenetriaminepentaacetic acid; CDTA, *trans*-1,2-diaminocyclohexane,*N,N,N',N'*-tetraacetic acid; Try, tryptophan; AA, anthranilic acid; XA, zanthurenic acid; KA, kynurenic acid; CySH, cysteine; Cys, cystine; His, histidine; Glu, glutamic acid.
7 1.8 mmol added per kilogram of diet.
8 5 g added per kilogram of diet.

From Nielsen, F. H. et al., *J. Nutr.*, 89, 35, 1966. With permission.

H. Chelates in Natural Materials

Nutritional studies over the years have shown that the addition of certain foodstuffs to a diet causes improved performance in experimental animals, and an important procedure has been the fractionation of the foodstuff to characterize the factor(s) responsible for the improved performance. This has resulted in the recognition of the vitamins, minerals, and other nutrients which we know today. Liver extract contains many known nutrients and has also been considered a source of unidentified growth factors. Scott and Ziegler[62] fed varying levels of Zn to chicks receiving a basal diet containing isolated soybean protein supplemented with 0, 1.5, or 3% of liver extract (Table 7). Growth was improved and the incidence of enlarged hocks reduced with the addition of graded levels of Zn to the diet. With the addition of 3% liver extract to the diet, less Zn was required for an optimum growth response and improvement in the condition of the hock joints. The response with 1.5% liver extract was intermediate between the 0 and 3% levels. The amount of Zn added by 3% liver extract was approximately 1.5 mg/kg, while the biological response to the addition of 3% liver extract was equivalent to approximately 60 mg/kg. It appears that the liver extract contains a factor(s) which improves the availability of Zn in the diet. Corn distillers dried solubles were also tested and found to have a similar, but less dramatic effect. This work shows that it is possible to obtain a biological response to the addition of a food material to a diet without that material contributing a known nutrient to the diet.

In studies with labeled chicken tissue, Janghorbani et al.[63] found that the absorption of Zn from intrinsically labeled meat was higher than from extrinsically labeled meat. In this instance, the Zn might be in a more absorbable complex form when it was released from the protein during digestion.

A number of low molecular weight compounds with metal-binding activity are present in plants. These are important since plants are intermediates in the overall movement of trace minerals through the food chain. Walker and Welch[64] fractionated soybeans grown in a nutrient solution containing ^{65}Zn. A molecular weight Zn- and Cu-binding compound was isolated and partially characterized. It did not contain S or P and did not appear to be similar to any of the common plant metabolites, such as citric acid. The availability of Zn from this compound is not known.

Ascorbic acid, which enhances the absorption of Fe, was studied for its effect on Zn absorption in humans;[65] Ascorbic acid (2 g) added to a bean gruel containing 108 mg of Zn failed to influence the plasma Zn level obtained with Zn alone, indicating that there is no effect of ascorbic acid on Zn absorption in the human subject.

A comparison of the effect of wine and ethanol upon Zn balance in humans was carried out by McDonald and Margen.[66] The young male subjects were fed a controlled diet and given 1 ℓ/day of Zinfandel wine, dealcoholized Zinfandel wine, an ethanol solution, or deionized water for four 18-day periods. There was increased absorption and, perhaps, decreased endogenous secretion of Zn during the wine and dealcoholized wine periods as compared to ethanol and deionized water. This effect was presumed to be due to the nonalcoholilc constituents of wine.

The ligands involved in Zn absorption in the rat were studied by Song and Adham.[67] The supernatant of cytoplasmic extract of rat small intestine was mixed with ^{65}Zn and chromatographed on Sephadex. The ^{65}Zn was located in four peaks and in the lowest molecular weight one was found to be associated with prostaglandin. In studies with isolated intestinal sacs, the addition of prostaglandin E_2 or prostaglandin extracted from rat small intestine resulted in a two- to threefold increase in the transfer of ^{65}Zn to the external medium. The results show that prostaglandin can chelate Zn and that it can enhance its transport across the intestinal mucosa.

Browning et al.[68] noted that the pathological signs of Zn and essential fatty acid deficiencies in rats are very similar and that the same symptoms may be induced by feeding toxic levels

Table 7
EFFECT OF Zn AND LIVER EXTRACT ON GROWTH OF CHICKS

		0% Liver extract			1.5% Liver extract			3.0% Liver extract		
Group	Zinc added[a] (mg/kg)	4-Week weight (g)	Feed/ gain	Incidence of enlarged hocks (%)	4-Week weight (g)	Feed/ gain	Incidence of enlarged hocks (%)	4-Week weight (g)	Feed/ gain	Incidence of enlarged hocks (%)
I	5	281[b]	1.99	73	356	1.86	73	410	1.74	70
	10	380	1.85	69	466	1.64	57	514	1.64	26
	15	457	1.67	46	524	1.62	30	556	1.63	8
II	60	524	1.60	5	545	1.58	5	560	1.56	3

Analysis of variance

Source of variation	Degrees of freedom	Mean square	F value[c]
Treatments	11	16,196	30
Group I (Zn deficient)	8	15,944	29
Zn	2	41,016	76
Liver extract	2	22,366	42
Zn × liver extract	4	197	NS
Group II (Zn adequate)	2	654	NS
Group I vs. Group II	1	49,289	92
Error	12	535	

[a] Added to soy protein basal diet C of Zeigler et al.
[b] Results are average of two lots of 15 male chicks per lot.
[c] Denotes significance at 0.01 level.

From Scott, M. L. and Zeigler, T. R., *J. Agric. Food Chem.*, 11, 123, 1963. With permission.

of aspirin, an inhibitor of prostaglandin synthesis. The possibility that impaired prostaglandin synthesis is the mechanism for the manifestation of typical deficiency symptoms is intriguing; however, in a study of the level of prostaglandin metabolites in the plasma of Zn-deficient rats, no difference could be detected which would indicate impaired prostaglandin synthesis. In similar studies with pregnant female rats, O'Dell et al.[69] found no decrease in prostaglandin synthesis in a Zn deficiency; in fact, there was a slight increase in plasma prostaglandin levels. It appears that prostaglandin synthesis is not the mechanims relating Zn to the development of the deficiency symptoms, but there may be some altered prostaglandin metabolism in a deficiency of Zn.

I. Human vs. Cow's Milk

There are two main observations which suggest that there is a difference in the availability of Zn from human and cow's milk. These are based on studies of the Zn content of hair, which led to studies of plasma Zn in infants and observations on acrodermatitis enteropathica.

Walravens and Hambidge[70] found that the plasma Zn levels in the infants in Colorado were lower than of normal adults, but correspond to levels for similar-aged American infants. On the other hand, reports on Swedish and German infants indicate that their plasma Zn levels are equal to those of older children and adults. The low Zn levels found in American infants could not be considered normal and raised the question of whether Zn is equally available from human milk and from cow's milk, on which the majority of American infant formulas were based. Further studies[71] showed that breast-fed infants have higher plasma Zn levels than those fed formula based upon cow's milk, even when supplemented with Zn.

Acrodermatitis enteropathica is a genetic disorder of infants characterized by alopecia, severe skin lesions, diarrhea, and early death. It was found to respond to treatment with the antifungal agent, diodoquin (5,7 di-iodo-8-hydroxyquinoline)

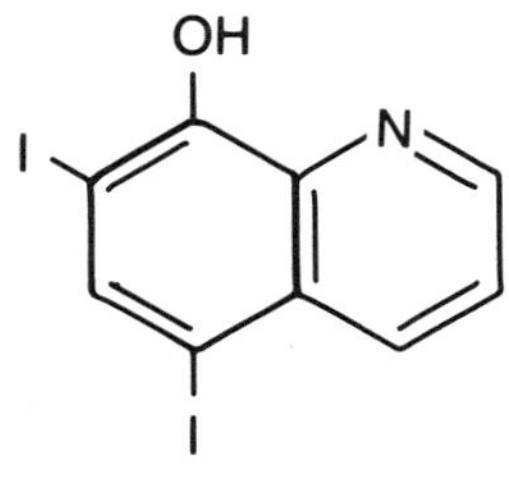

Diodoquin or Diiodohydroxyquin

which is a Zn-binding ligand.[72] Whole body measurements of patients[73] after administration of ^{65}Zn showed that the low serum Zn concentrations were due to reduced absorption of Zn in the gut and not by an increased excretion of Zn into the feces, urine, or sweat. Jejunal biopsies from patients with acrodermatitis enteropathica showed less accumulation of ^{65}Zn than their controls, suggesting that uptake of Zn by the enterocytes is the primary abnormality responsible for the disorder. Administering large quantities of Zn as in the treatment of the disorder by oral Zn therapy[74] is able to overcome the defect.

Breast feeding or treatment with human milk prevents the symptoms of acrodermatitis enteropathica and maintains normal plasma Zn levels, although the Zn content of human milk after the 1st month of lactation is lower than of cow's milk. Casey et al.[75] determined the availability of Zn for young adult women using a standard Zn loading test. The subjects consumed 25 mg of Zn with milk or formula to provide 5 g of protein (Figure 7). Blood samples were taken initially and at 30-min intervals for 3 hr after the administration of the Zn. The response was measured as the increase in plasma Zn above the baseline over the

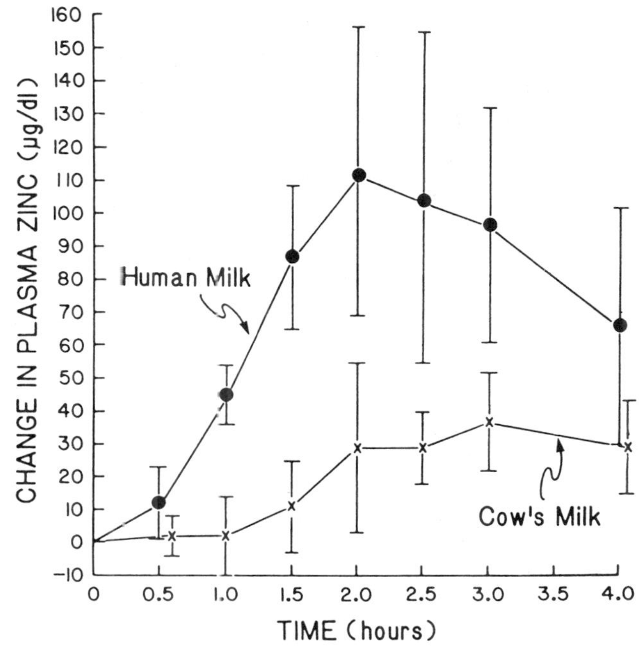

FIGURE 7. Changes in plasma Zn concentration after ingestion of 25 mg of Zn with human milk (five subjects) and cow's milk (seven subjects). Points represent means ± SD variation from baseline value. (From Casey, C. E. et al., *Pediatrics*, 68, 394, 1981. Copyright American Academy of Pediatrics 1981. With permission.)

3-hr period. The response with cow's milk was approximately one quarter that obtained with human milk. Zn from an infant formula based upon cow's milk had Zn availability very similar to that with cow's milk alone (Table 8), while those based upon soybean protein and casein hydrolysates had even poorer availability. These differences in Zn availability between human and cow's milk are greater than have been reported in rat studies, but point to the importance of the Zn availability problem in infant nutrition.

The reason for the difference in availability of Zn from human and cow's milk has received considerable attention in recent years. An attractive hypothesis is that the Zn is bound in different ways in these two types of milk and that the nature of the complex would influence the biological availability of the metal. In gel filtration studies, Eckhert et al.[76] found that a significant fraction of the Zn in human milk was associated with a low molecular weight Zn-binding ligand, while Zn in cow's milk was bound almost exclusively to high molecular weight compounds. By modifying their gel filtration technique, Lonnerdal et al.[77] were able to show that Zn in human milk was complexed only with citrate. Synthetic Zn citrate applied to the gel filtration or ion-exchange columns eluted in the same position as the Zn-containing compound from human milk.[78] The concentration of citrate in cow's milk was higher than in human milk, but there was very little Zn association with it. It is probable that the high molecular weight material with which Zn is associated in cow's milk is casein. It is approximately ten times more abundant in cow's milk than human milk and has high Zn-binding activity, presumably due to its phosphoserine groups. In addition, casein is not as well digested by infants as human milk proteins, which would prolong the binding activity in the GI tract. Hurley and Lonnerdal[78] have proposed that the citrate with which the Zn is bound in human milk is important in making the metal highly available.

In other studies on the same problem, Evans and Johnson[79] proposed that the low molecular

Table 8
PLASMA Zn RESPONSE: Zn (25 mg) TAKEN WITH HUMAN MILK, COW'S MILK, AND INFANT FORMULAS

Product	No. of subjects	Plasma Zn response[a]
Human milk	5	200.6 ± 56.8
2% Cows milk	7	50.9 ± 37.6[b,c]
Formulas		
Cow's milk-based	6	56.9 ± 23.2[b,d]
Soy-based	7	33.9 ± 17.2[b,c]
Casein hydrolysate A	3	35.3 ± 17.1[b]
Casein hydrolysate B	5	15.9 ± 7.8[b]

[a] Area under curve of change in plasma Zn from baseline value plotted against time over 3 hr.
[b] Significantly less ($p < 0.005$) than human milk.
[c] Significantly greater ($p < 0.05$) than casein hydrolysate B.
[d] Significantly greater than soy-based ($p < 0.05$) and casein hydrolysate B ($p < 0.01$).

From Casey, C. E. et al., *Pediatrics*, 68, 394, 1981. With permission.

weight ligand in human milk is picolinic acid. They reported that the concentration of picolinic acid in one sample of human milk was 308 μmol while the concentration in one brand of processed cow's milk was 20 μmol. It could not be detected in another sample of cow's milk and in four infant formulas. They suggested that the higher biological availability of Zn from human milk than cow's milk is due to the presence of picolinic acid in the human milk. They also suggested an abnormality in tryptophan metabolism in acrodermatitis enteropathica patients, limiting the synthesis of picolinic acid.[80,81] The presence of picolinic acid in the pancreatic enzyme extract, the fact that ligation of the pancreatic duct impairs Zn absorption, and the therapeutic response of infants with acrodermatitis enteropathica to supplements of picolinic acid are cited as support of the importance of picolinic acid in Zn absorption. Evans[80] proposed that the exocrine cells of the pancreas produce picolinic acid, which is secreted into the lumen of the intestine where it coordinates with Zn to form a complex which facilitates the passage of Zn through the luminal membrane. Whether the picolinic acid complex has an important biological role has been questioned because of the low level found in milk, and its presence in pancreatic juice is controversial.[78,82]

Cousins and Smith[83] found the total Zn content of fat-free human and bovine milk to be very similar (3.5 and 3.6 μg/mℓ, respectively) while the protein content was 5.3 and 29.0 mg/mℓ, respectively. By chromatography, it was found that about 10% of the Zn in human milk was associated with low molecular weight fractions. When Zn was added to the milk in vitro, the presence of low molecular weight fractions could be noted in the milk from both sources, but to a greater extent in the human milk. These results are consistent with the idea that the much higher content of casein with its high Zn-binding affinity in cow's milk accounts for the high proportion of Zn bound to high molecular weight sites in this milk. In human milk the high molecular weight binding sites are saturated and a higher proportion of the Zn is associated with low molecular weight fractions.

REFERENCES

1. **Cousins, R. J.,** Mechanism of zinc absorption, in *Current Topics in Nutrition and Disease,* Vol. 6, Prasad, A. S., Ed., Alan R. Liss, New York, 1982, chap. 5.
2. **Kirchgessner, M. and Wiegand, E.,** Zinc absorption and excretion in relation to nutrition, in *Metal Ions In Biological Systems,* Vol. 15, Sigel, H., Ed., Marcel Dekker, New York, 1983, 319.
3. **Solomons, N. W.,** Biological availability of zinc in humans, *Am. J. Clin. Nutr.,* 35, 1048, 1982.
4. **DiSilvestro, R. A. and Cousins, R. J.,** Physiological ligands for copper and zinc, in *Annual Review of Nutrition,* Darby, W. J., Broquist, H. P., and Olson, R. E., Eds., Annual Reviews, Palo Alto, Calif., 1983, 261.
5. **Starcher, B. C., Glauber, J. G., and Madaras, J. G.,** Zinc absorption and its relationship to intestinal metallothionein, *J. Nutr.,* 110, 1391, 1980.
6. **Kowarski, S., Blair-Stanek, C. S., and Schacter, D.,** Active transport of zinc and identification of zinc-binding protein in rat jejunal mucosa, *Am. J. Physiol.,* 226, 401, 1974.
7. **Davies, N. T.,** Studies on the absorption of zinc by rat intestine, *Br. J. Nutr.,* 43, 189, 1980.
8. **O'Dell, B. L. and Savage, J. E.,** Potassium, zinc and distillers dried solubles as supplements to a purified diet, *Poult. Sci.,* 36, 459, 1957.
9. **Morrison, A. B. and Sarett, H. P.,** Studies of zinc deficiency in the chick, *J. Nutr.,* 65, 267, 1958.
10. **Moeller, M. W. and Scott, H. M.,** Studies with purified diets. III. Zinc requirements, *Poult. Sci.,* 37, 1227, 1958.
11. **Norris, L. C., Leach, R. M., Jr., and Zeigler, T. R.,** Recent research on the mineral requirements of poultry, *Proc. Distill. Feed Conf.,* 13, 63, 1958.
12. **Kratzer, F. H., Allred, J. B., Davis, P. N., Marshall, B. J., and Vohra, P.,** The effect of autoclaving soybean protein and the addition of EDTA on the availability of dietary zinc for poults, *J. Nutr.,* 68, 313, 1959.
13. **Forbes, R. M. and Parker, H. M.,** Biological availability of zinc in and as influenced by whole fat soy flour in rat diets, *Nutr. Rep. Int.,* 15, 681, 1977.
14. **O'Dell, B. L.,** Effect of soy protein on trace mineral availability, in *Soybean in Human Nutrition,* Wilcke, H., Hopkins, D., and Waggle, D., Eds., Academic Press, New York, 1979, 198.
15. **Momcilovic, B., Bartholomeus, B., Giroux, A., and Shah, B. G.,** Bioavailability of zinc in milk and soy protein-based infant formulas, *J. Nutr.,* 106, 913, 1976.
16. **Forbes, R. M. and Parker, H. M.,** Biological availability of zinc in and as influenced by whole fat soy flour in rat diets, *Nutr. Rep. Int.,* 15, 681, 1977.
17. **O'Dell, B. L., Burpo, C. E., and Savage, J. E.,** Evaluation of zinc availability in foodstuffs of plants and animal origin, *J. Nutr.,* 102, 653, 1972.
18. **O'Dell, B. L. and Savage, J. E.,** Effect of phytic acid on zinc availability, *Proc. Soc. Exp. Biol. Med.,* 103, 304, 1960.
19. **Oberlease, D., Muhrer, M. E., and O'Dell, B. L.,** Effects of phytic acid on zinc availability and parakeratosis in swine, *J. Anim. Sci.,* 21, 57, 1962.
20. **Oberlease, D., Muhrer, M. E., and O'Dell, B. L.,** Dietary metal-complexing agents and zinc availability in the rat, *J. Nutr.,* 90, 56, 1966.
21. **Likuski, H. J. and Forbes, R. M.,** Effects of calcium and phytic acid on the utilization of dietary zinc, *J. Nutr.,* 85, 230, 1965.
22. **Rheinhold, J. G., Nasr, K., Lahimgarzadeh, A., and Hedayati, H.,** Effects of purified phytate and phytate-rich bread upon metabolism of zinc, calcium, phosphorus and nitrogen in man, *Lancet,* 1, 283, 1973.
23. **Rackis, J. J., McGhee, J. E., Honig, D. H., and Booth, A. N.,** Processing soybeans into foods: selected aspects of nutrition and flavor, *J. Am. Oil Chem. Soc.,* 52, 249A, 1975.
24. **Ketelsen, S. M., Stuart, M. A., Weaver, C. M., Forbes, R. M., and Erdman, J. W., Jr.,** Bioavailability of zinc to rats from defatted soy flour, acid-precipitated soy concentrate and neutralized soy concentrate as determined by intrinsic and extrinsic labeling techniques, *J. Nutr.,* 114, 536, 1984.
25. **Lease, J. G., Barnett, B. D., Lease, E. J., and Turk, D. E.,** The biological availability to the chick of zinc in a sesame meal ration, *J. Nutr.,* 72, 66, 1960.
26. **Rheinhold, J. G., Hedayate, H., Lahimgarzadeh, A., and Nasr, K.,** Zinc, calcium, phosphorus and nitrogen balance of Iranian villagers following a change from pytate-rich to phytate-poor diet, *Ecol. Food Nutr.,* 2, 157, 1973.
27. **Davies, N. T., Hristic, V., and Flett, A. A.,** Phytate rather than fiber in bran as the major determinant of zinc availability to rats, *Nutr. Rep. Int.,* 15, 207, 1977.
28. **Forbes, R. M., Parker, H. M., and Erdman, J. W., Jr.,** Effects of dietary phytate, calcium and magnesium levels on zinc bioavailability to rats, *J. Nutr.,* 114, 1421, 1984.
29. **Gaf, E. and Eaton, J.,** Effects of phytate on mineral bioavailability in mice, *J. Nutr.,* 114, 1192, 1984.

30. **Reinhold, J. G., Faradji, B., Abadi, P., and Ismail-Beigi, F.,** Binding of zinc to fiber and other solids of wholemeal bread, in *Trace Elements in Human Health and Disease,* Vol. 1, Prasad, A. S., Ed., Academic Press, New York, 1976, 163.
31. **Sandstrom, B., Arvidsson, B., Cederblad, A., and Bjorn-Rasmussen, E.,** Zinc absorption from composite meals. I. Significance of wheat extraction rate, zinc, calcium and protein content in meals based on bread, *Am. J. Clin. Nutr.,* 33, 739, 1980.
32. **Reinhold, J. G., Faraji, B., Abadi, P., and Ismail-Beigi, F.,** Decreased absorption of calcium, magnesium, zinc and phosphorus by humans due to increased fiber and phosphorus consumption of wheat bread, *J. Nutr.,* 106, 493, 1976.
33. **Ismail-Beigi, F., Reinhold, J. G., Faraji, B., and Abadi, P.,** Effects of cellulose added to diets of low and high fiber content upon the metabolism of calcium, magnesium, zinc and phosphorus by man, *J. Nutr.,* 107, 510, 1977.
34. **Guthrie, B. E. and Robinson, J. F.,** Zinc balance studies during wheat bran supplementation, *Fed. Proc. Fed. Am. Soc. Exp. Biol.,* 37, 254, 1978.
35. **Sandstead, H., Klevay, L., Munoz, J., Jaacob, R., Logan, G., Jr., Dintzis, F., Inglett, G., and Shuey, W.,** Human zinc requirements: effects of dietary fiber on zinc metabolism, *Fed. Proc. Fed. Am. Exp. Biol.,* 37, 254, 1978.
36. **Smith, J. C., Jr., Morris, E. R., and Ellis, R.,** Zinc: requirements, bioavailabilities, and recommended dietary allowances, in *Zinc Deficiency in Human Subjects,* Prasad, A. S., Cavdar, A. O., Brewer, G. J., and Aggett, P. J., Eds., Alan R. Liss, New York, 1983, 147.
37. **Tasi, R. C. Y. and Lei, K. Y.,** Dietary cellulose zinc and copper: effect on tissue levels of trace minerals in the rat, *J. Nutr.,* 109, 1117, 1979.
38. **Harmuth-Hoene, A. E. and Schelenz, R.,** Effect of dietary fiber on mineral absorption in growing rats, *J. Nutr.,* 110, 1774, 1980.
39. **Morris, E. R. and Ellis, R.,** Bioavailability to rats of iron and zinc in wheat bran: response to low-phytate bran and effect of the phytate/zinc molar ratio, *J. Nutr.,* 110, 2000, 1980.
40. **Lantzsch, H. J., Scheuermann, S. E., and Marschner, H.,** Improvement of zinc availability in cereal grains, in *Trace Element Metabolism in Man and Animals,* Proc. 4th Int. Symp. (TEMA 4), Gawthorne, J. M., McC. Howell, J., and White, C. L., Eds., Springer-Verlag, Basel, 1982, 107.
41. **Sandstrom, B. and Cederblad, A.,** Zinc absorption from composite meals. II. Influence of the main protein source, *Am. J. Clin. Nutr.,* 33, 1778, 1980.
42. **Maga, J. A.,** Phytate: its chemistry, occurrence, food interactions, nutritional significance and methods of analysis, *J. Agric. Food Chem.,* 30, 1, 1982.
43. **Allred, J. B., Kratzer, F. H., and Porter, J. W. G.,** Some factors affecting the *in vitro* binding of zinc by isolated soybean protein and by α-casein, *Br. J. Nutr.,* 18, 575, 1964.
44. **Lease, J. G. and Williams, W. P., Jr.,** Availability of zinc and comparison of *in vitro* and *in vivo* zinc uptake of certain oil seed meals, *Poult. Sci.,* 46, 233, 1967.
45. **Shah, B. G., Giroux, A., Belonge, B., and Jones, J. D.,** Optimal level of zinc supplementation for young rats fed rapeseed protein concentrate, *J. Agric. Food Chem.,* 27, 387, 1979.
46. **Pecoud, A. and Schelling, D. P.,** The effect of foodstuffs on the absorption of zinc sulfate, *Clin. Pharmacol. Ther.,* 17, 469, 1975.
47. **Shah, B. G., Meranger, J. C., Belonje, B., and Deshmukh, D. S.,** Zinc retention in young rats fed increasing levels of high-zinc oysters, *J. Fish. Res. Board Can.,* 28, 843, 1971.
48. **Kratzer, F. H. and Starcher, B.,** Quantitative relation of EDTA to availability of zinc for turkey poults, *Proc. Soc. Exp. Biol. Med.,* 113, 424, 1963.
49. **Koike, T. L., Vohra, P., and Kratzer, F. H.,** Intestinal absorption of zinc or calcium-EDTA complexes in chickens, *Proc. Soc. Exp. Biol. Med.,* 117, 483, 1964.
50. **Foreman, H.,** *Metal Binding in Medicine,* Seven, M. J. and Johnson, L. S., Eds., J. B. Lippincott, New York, 1960, 82.
51. **Vohra, P. and Gonzales, N.,** The effect of EDTA on the preferential intestinal absorption of zinc than manganese in turkey poults, *Poult. Sci.,* 48, 1509, 1969.
52. **Forbes, R. M.,** Excretory patterns and bone depositin of zinc, calcium and magnesium in the rat as influenced by zinc deficiency, EDTA and lactose, *J. Nutr.,* 74, 194, 1961.
53. **Oberleas, D., Muhrer, M. E., and O'Dell, B. L.,** Dietary metal-complexing agents and zinc availability in the rat, *J. Nutr.,* 90, 56, 1966.
54. **Likuski, H. J. A. and Forbes, R. M.,** Effect of phytic acid on the availability of zinc in amino acid and casein diets fed to chicks, *J. Nutr.,* 84, 145, 1964.
55. **Suso, F. A. and Edwards, H. M., Jr.,** A study of techniques of measuring ^{65}Zn absorption and biological half-life in the chicken, *Poult. Sci.,* 47, 991, 1968.
56. **Suso, F. A. and Edwards, H. M., Jr.,** Influence of various chelating agents on absorption of ^{60}Co, ^{59}Mn and ^{65}Zn by chickens, *Poult. Sci.,* 47, 1417, 1968.

57. **Vohra, P. and Kratzer, F. H.,** Influence of various chelating agents on the availability of zinc, *J. Nutr.*, 82, 249, 1964.
58. **Nielsen, F. H., Sunde, M. L., and Hoekstra, W. G.,** Effects of some dietary synthetic and natural chelating agents on the zinc-deficiency syndrome in the chick, *J. Nutr.*, 89, 35, 1966.
59. **Oestricher, P. and Cousins, R. J.,** Influence of intraluminal constituents on zinc absorption by isolated, vascularly perfused rat intestine, *J. Nutr.*, 112, 1978, 1982.
60. **Welch, R. M., House, W. A., and van Campen, D.,** Effects of oxalic acid on availability of zinc from spinach leaves and zinc sulfate in rats, *J. Nutr.*, 107, 929, 1977.
61. **Weismann, K., Knudsen, L., and Hoyer, H.,** Phenytoin increases [65]zinc absorption in the rat, *J. Invest. Dermatol.*, 71, 396, 1978.
62. **Scott, M. L. and Zeigler, T. R.,** Chelation in nutrition. Evidence for natural chelates which aid in the utilization of zinc by chicks, *J. Agric. Food Chem.*, 11, 123, 1963.
63. **Janghorbani, M., Istfan, N. W., Pagounas, J. O., Steinke, F. H., and Young, V. R.,** Absorption of dietary zinc in man: comparison of intrinsic and extrinsic labels using a triple stable isotope method, *Am. J. Clin. Nutr.*, 36, 537, 1982.
64. **Walker, C. D. and Welch, R. M.,** Low molecular weight metal compounds in edible plants, in *Trace Element Metabolism in Man and Animals*, Proc. 4th Int. Symp. (TEMA 4), Gawthorne, J. M., McC. Howell, J., and White, C. L., Eds., Springer-Verlag, Basel, 1982, 155.
65. **Solomons, N. W., Jacob, R. A., Pineda, O., and Viteri, F. E.,** Studies on the bioavailability of zinc in man. III. Effect of ascorbic acid on zinc absorption, *Am. J. Clin. Nutr.*, 32, 2495, 1979.
66. **McDonald, J. T. and Margen, S.,** Wine versus ethanol in human nutrition. IV. Zinc balance, *Am. J. Clin. Nutr.*, 33, 1096, 1980.
67. **Song, M. K. and Adham, N. F.,** Role of prostaglandin E_2 in zinc absorption, *Am. J. Physiol.*, 234, 99, 1978.
68. **Browning, J. D., Reeves, P. G., and O'Dell, B. L.,** Effect of zinc deficiency and food restriction on the plasma levels of prostaglandin metabolites in male rats, *J. Nutr.*, 113, 755, 1983.
69. **O'Dell, B. L., Browning, J. D., and Reeves, P. G.,** Plasma levels of prostaglandin metabolites in zinc-deficient female rats near term, *J. Nutr.*, 113, 760, 1983.
70. **Walravens, P. A. and Hambidge, K. M.,** Growth of infants fed a zinc supplemented formula, *Am. J. Clin. Nutr.*, 29, 1114, 1976.
71. **Hambidge, K. M., Walravens, P. A., Casey, C. E., Brown, R. M., and Bender, J.,** Plasma zinc concentration of breast-fed infants, *Pediatrics*, 94, 607, 1979.
72. **Aggett, P. J., Delves, H. T., Harries, J. T., and Bangham, A. D.,** The possible role of diodoquin as a zinc ionophore in the treatment of acrodermatitis enteropathica, *Biochem. Biophys. Res. Commun.*, 87, 513, 1979.
73. **Lombeck, I., Schnippering, H. C., Ritzl, F., Feinendegen, L. E., and Bremer, H. J.,** Absorption of zinc in acrodermatitis enteropathica, *Lancet*, 1, 855, 1975.
74. **Moynahan, E. J.,** Acrodermatitis enteropathica: a lethal inherited human zinc-deficiency disorder, *Lancet*, ii, 399, 1974.
75. **Casey, C. E., Walravens, P. A., and Hambidge, K. M.,** Availability of zinc: loading test with human milk, cow's milk and infant formulas, *Pediatrics*, 68, 394, 1981.
76. **Eckhert, C. D., Sloan, M. V., Duncan, J. R., and Hurley, L. S.,** Zinc binding: a difference between human and bovine milk, *Science*, 195, 789, 1977.
77. **Lonnerdal, B., Keen, C. L., Sloan, M. V., and Hurley, L. S.,** Molecular localization of zinc in rat milk and neonatael intestine, *J. Nutr.*, 10, 2414, 1980.
78. **Hurley, L. S. and Lonnerdal, B.,** Zinc binding in human milk: citrate versus picolinate, *Nutr. Rev.*, 40, 65, 1982.
79. **Evans, G. W. and Johnson, P. E.,** Characterization and quantitation of a zinc-binding ligand in human milk, *Pediatr. Res.*, 14, 876, 1980.
80. **Evans, G. W.,** Normal and abnormal zinc absorption in man and animals: the tryptophan connection, *Nutr. Rev.*, 38, 137, 1980.
81. **Sandstead, H. H. and Evans. G. W.,** Zinc, in *Present Knowledge in Nutrition*, 5th ed., Olson, R. E., Broquist, H. P., Chichester, C. O., Darby, W. J., Kolbye, A. C., Jr., and Stalvey, R. M., Eds., The Nutrition Foundation, Washington, D.C., 1984, chap. 33.
82. **Hurley, L. S. and Lonnerdal, B.,** Picolinic acid as a zinc-binding ligand in human milk: an unconvincing case, *Pediatr. Res.*, 15, 166, 1981.
83. **Cousins, R. J. and Smith, K. T.,** Zinc-binding properties of bovine and human milk in vitro: influence of changes in zinc content, *Am. J. Clin. Nutr.*, 33, 1083, 1980.

Chapter 7

IRON

I. INTRODUCTION

Both divalent and trivalent Fe have very high affinities to combine with proteins and other complexing agents at near neutral pH and little free Fe is found in the animal. In the blood, transferrin with 74,000 mol sat, has a high affinity for both Fe^{2+} and Fe^{3+}. A loading of Fe above the capacity of transferrin causes Fe toxicity. In tissues, apoferritin combines with ferric ions to an extent of about 25% of its weight and is found in hemosiderin combined with OH^-, sugars, and protein to an extent of over 30%. The passage of Fe through membranes is in the divalent form, presumably complexed with an organic ligand. When Fe^{2+} appears in the blood it is oxidized to Fe^{3+} by ceruloplasmin and is bound to transferrin to be transported to tissues.[1]

The topic of absorption of Fe has been reviewed recently by Hallberg,[2] Morris,[3] and Finch and Cook.[4] Food Fe can be classified into heme Fe from hemoglobin and myoglobin and nonheme Fe largely supplied by cereals, fruit, and vegetables. It is estimated that 40% of the Fe in animal tissue is in the heme form. Diets low in animal tissue contain negligible amounts of heme Fe and nonheme Fe is the main source.

Heme Fe is thought to be absorbed intact and freed from the heme complex in the mucosal cell as is illustrated in Figure 1 from Finch and Cook.[4] Heme Fe is utilized with relatively little influence from other dietary factors. Nonheme Fe, on the other hand, is less absorbed and its absorption is influenced by several enhancing and inhibiting factors.

In addition to dietary factors which influence Fe utilization, the Fe status of the individual greatly influences the utilization of this element. For example, Fe absorption is several times more efficient in a severely deficient individual than in an Fe-replete individual. The Fe status influences the utilization of nonheme Fe more than heme, but there is a definite effect of Fe status on heme Fe utilization at high levels of heme Fe as one would have with the consumption of blood sausage.[5] Meat is the only food source known to influence heme Fe absorption[6] and it can be shown that the absorption of Fe from heme is improved by the addition of meat to a diet.

II. AVAILABILITY OF Fe FROM DIFFERENT Fe COMPOUNDS

The availability of Fe from its various salts for human subjects may be inferred from studies by Brise and Hallberg.[7] They found that no Fe compound was better than slightly soluble, quite dissociated ferrous compounds. The citrate, tartrate, and pyrophosphate ferrous salts, in which part of the Fe is complex bound, had reduced availability. All of the ferric salts had low absorption values.

In hemoglobin repletion studies with chicks and rats, Fritz et al.[8] determined the availability of Fe from a number of compounds as shown in Table 1. Again, ferrous sulfate was well utilized and was used as the standard to report the relative value of the other compounds. In general, ferric compounds were less effectively utilized, although the Fe salts of EDTA, ammonium citrate, and choline citrate were utilized as well as the ferrous sulfate.

III. EFFECT OF VARIOUS CHELATING AGENTS ON THE ABSORPTION OF NONHEME Fe

A. Ascorbic Acid

It is well established that ascorbic acid enhances the absorption of Fe in certain species

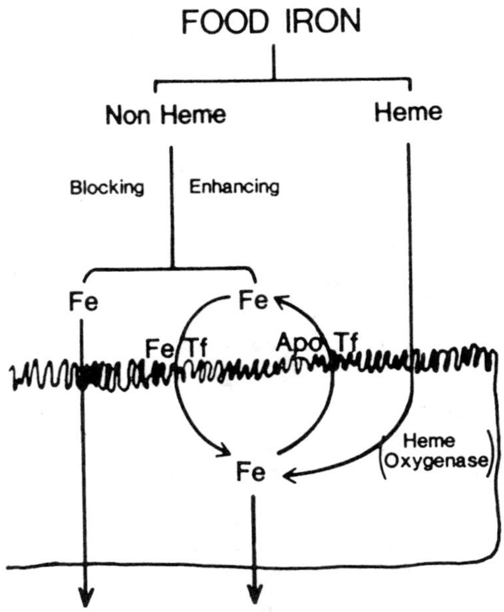

FIGURE 1. The process of Fe absorption is illustrated diagrammatically. Fe is absorbed as heme and nonheme Fe. Heme Fe is absorbed intact and processed within the mucosal cell. The absorption of nonheme Fe is effected by blocking and enhancing substances within the diet. Its Fe may be absorbed either chelated to the protein mucosal transferrin or may pass through the mucosa by diffusion. (From Finch, C. A. and Cook, J. D., *Am. J. Clin. Nutr.*, 39, 471, 1984. With permission.)

such as rats[9] and pigs,[10] but both Hill and Starcher[11] and Fritz et al.[8] found no effect of ascorbic acid on Fe utilization in the chick at levels of 0.02 and 0.1% of the diet. Miski and Kratzer[12] also investigated the problem and observed increased body retention of Fe with 0.1% ascorbic acid and increased hemoglobin and body retention with 0.4% ascorbic acid. It appears that the level of ascorbic acid used is critical for its effect on Fe absorption.

By the use of a double isotope method, Brise and Hallberg[13] showed that orally administered ascorbic acid increased the absorption of Fe from ferrous sulfate in human subjects. Low levels were ineffective, but 200 mg of ascorbic acid with 30 mg of ferrous Fe proved to be an adequate level orally. When the ascorbic acid was administered intravenously and Fe orally, there was no effect on either Fe absorption or the turnover of plasma Fe. It was thus concluded that the site of action of the ascorbic acid on Fe absorption was the GI lumen and that its action was probably due to its reducing action in preventing or delaying the formation of less dissociated and insoluble ferric compounds. Lee et al.[14] confirmed the effect of ascorbic acid on Fe absorption in humans. By measuring serum Fe levels, they showed that 1 g of ascorbic acid administered with 200 mg of ferrous sulfate increased serum Fe levels while 50 mg of ascorbic acid was ineffective. Grebe et al.[15] observed that ascorbic acid enhanced the absorption of therapeutic doses of Fe, with larger amounts needed with ferrous salts than with ferric salts.

Fe absorption from isolated intestinal segments was studied in rats by Hungerford and Linder.[16] (See Figure 2.) When the ferric chloride solution at pH 2 was placed in the intestinal segment, the pH rose rapidly, reaching pH 4 (at which Fe^{3+} is insoluble) in about 5 min and pH 6.8 by 10 to 15 min. Without the addition of ascorbic acid, Fe uptake from the segment ceased after about 5 min, but with the addition of 2 m*M* ascorbate, Fe uptake continued for about 40 min and the total amount of Fe taken up was doubled. The pH within the intestinal segment increased as rapidly in the presence of ascorbic acid as in the other

Table 1
RELATIVE BIOLOGICAL VALUE OF Fe FROM VARIOUS DIETARY SOURCES

Fe source	No. samples	Relative biological value: Average	Relative biological value: Range
Fe Compounds			
EDTA, dihydrogen ferrous salt	1	99	97—100
Ferric ammonium citrate	1	107	98—115
Ferric choline citrate	1	102	
Ferric chloride	1	44	26—67
Ferric citrate	1	73	70—76
Ferric glycerophosphate	1	93	86—100
Ferric pyrophosphate	1	45	38—52
Ferric orthophosphate	4	14	7—32
Ferric oxide	1	4	0—6
Ferric sulfate	1	83	65—100
Ferrous ammonium sulfate	1	99	99—100
Ferrous carbonate	5	2	0—6
Ferrous chloride	1	98	
Ferrous fumarate	1	95	71—133
Ferrous gluconate	1	97	
Ferrous sulfate ($FeSO_4{\cdot}7H_2O$)	1	100	
Ferrous sulfate, anhydrous	1	100	
Ferrous sulfate, feed grade	1	100	
Ferrous tartrate	1	77	70—83
Reduced iron	6	37	8—66
Sodium iron pyrophosphate	3	14	2—23

From Fritz, J. C. et al. *J. Agric. Food Chem.*, 18, 647, 1970. With permission.

segment without ascorbic acid, so the effect on absorption could not be attributed to increasing the acidity but to keeping the Fe in a soluble and available form.

The effect of ascorbic acid on Fe and Cu utilization was studied in the guinea pig, a specie which requires this vitamin.[17] Cu absorption and utilization, as shown by serum Cu and ceruloplasmin activity, were increased in a deficiency of ascorbic acid and decreased by high-level supplementation. This is consistent with other species in which high-levels of ascorbic acid aggravate a Cu deficiency. Fe absorption was also increased by 250 mg/kg of ascorbic acid, consistent with animals not requiring this vitamin.

Brise and Hallberg[18] noted that the utilization of Fe was greater from a solution of ferrous succinate than from a solution of ferrous sulfate. This prompted a study to determine whether succinate ions themselves might influence Fe absorption. It was found that succinic acid increased the absorption of Fe when added to the oral Fe dose given to human subjects. The effect was independent of ascorbic acid since each caused a specific increase in the presence of the other. Succinic acid did not affect Fe turnover or intestinal motility. In contrast to ascorbic acid, succinic acid administered intravenously also increased Fe absorption. It thus appears that succinic acid affects absorption by increasing the intracellular mucosal metabolism and that the mechanism is not primarily one of chelation. One would expect that the transport of metals other than Fe would also be influenced by succinic acid.

B. Carbohydrates

Amine and Hegsted[19] found that many dietary factors influenced the absorption of ^{59}Fe

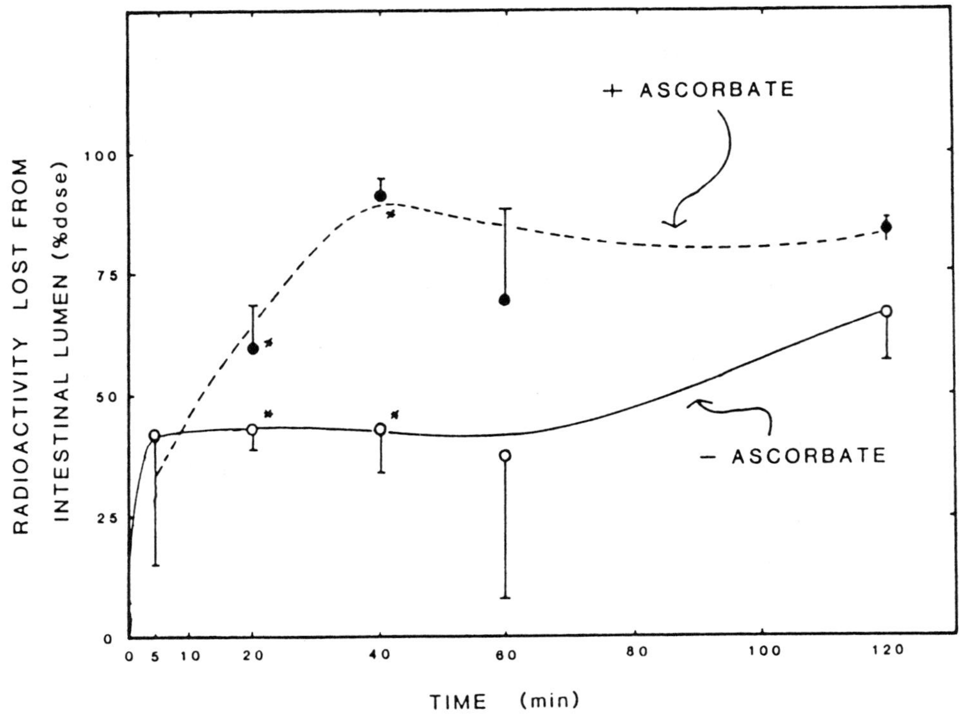

FIGURE 2. Time course of the uptake of ^{59}Fe from the intestinal lumen when administered in a pH 2 solution. Samples (0.5 mℓ) of $^{59}FeCl_3$, in 0.01 *M* HCl (3.855 ⓣⓜ 10^6 cpm) were placed in tied-off duodenal segments. Mean values for loss of radioactivity from the lumen are plotted against time for rats given the Fe solution without (solid line) or with (dashed line) added ascorbic acid. The number of rats in each group ranged from two to six. The bars represent 1 SD. (From Hungerford, D. M., Jr. and Linder, M. C., *J. Nutr.*, 113, 2615, 1983. With permission.)

Table 2
ESTIMATES OF THE RELATIVE AVAILABILITY OF DIFFERENT Fe SOURCES FED WITH DIFFERENT CARBOHYDRATES

Fe source	Starch (%)	Sucrose (%)	Lactose (%)
Ferrous sulfate	100.0	100.0	100.0
Reduced Fe	11.4 ± 6.0[a]	14.5 ± 5.6	30.9 ± 4.8
Sodium Fe pyrophosphate	23.7 ± 3.0	30.9 ± 5.2	23.8 ± 3.0
Ferric orthophosphate	33.9 ± 3.8	49.2 ± 7.6	27.9 ± 3.6

[a] Mean ± standard error.

From Amine, E. K. and Hegsted, D. M., *J. Agric. Food Chem.*, 23, 204, 1975. With permission.

in rats when they were fed with ferric ammonium sulfate. Significant differences in the amount were noted with different sources of carbohydrate with the percentage of Fe retained over a 9-day period greatest with lactose and decreasing with lactose plus starch, sucrose, glucose, and starch. In more extended studies, as shown in Table 2, the same workers[20] found a difference in Fe retention with various carbohydrates, depending upon the form of

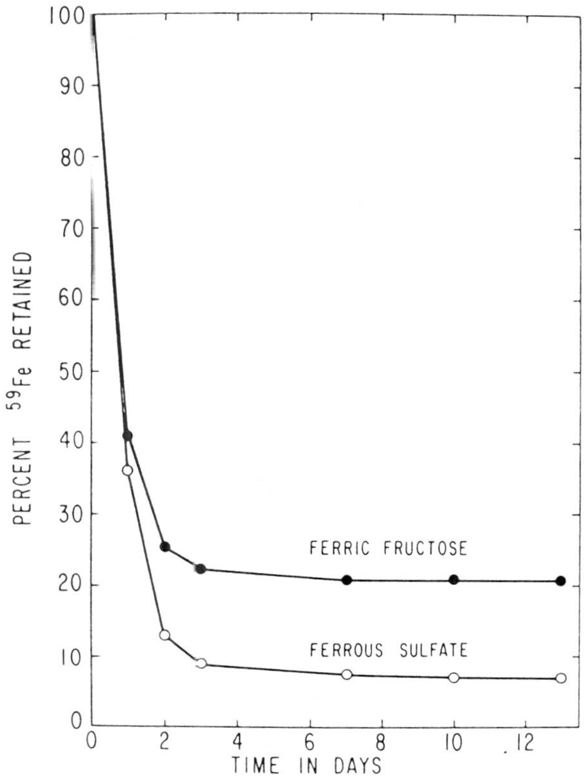

FIGURE 3. Retention of radioiron from a single oral dose of ferrous sulfate or ferric fructose as a function of time. The animals were given 1.0 mℓ of a solution containing either 250 m*M* ferrous sulfate (14 μg Fe) or 500 μ*M* iron-1 *M* fructose (28 μg Fe). Each datum point represents the average of six animals. (From Christopher, J. P. et al., *Trace Element Metabolism in Man*, Vol. 2, Proc. 2nd Int. Symp. Trace Element Metab. Anim., Madison, Wisc., 1973, 133. With permission.)

the Fe used. Lactose was particularly effective in improving the availability of reduced Fe, but was no more effective than starch and less effective than sucrose in influencing Fe absorption from sodium iron pyrophosphate and ferric orthophosphate, respectively. Sucrose promoted the best availability of Fe from ferric orthophosphate.

Considerable work has been done on the availability of Fe from iron fructose chelates.[21,22] Fructose has the ability to form a soluble complex with Fe(III), which can be readily absorbed. In a study with guinea pigs which were supplied ^{59}Fe as ferric fructate,[21] retention of Fe (Figure 3) was over twice as great as from ferrous sulfate at 10 to 12 days as determined by whole body counting. Not only is the complex readily absorbed, but the Fe is readily released so that it is used by the animal. The ferric fructose chelate was also shown to enhance Fe uptake in man.[21] In a double-labeled experiment in which ratios of fructose to Fe(III) of 20:1 and 100:1 were used. Fe absorption was about twice that of ferrous sulfate alone. The fructose level was important since a 2.5:1 ratio caused poorer utilization than the control.

C. Fiber and Phytate

Since fiber and phytate occur together in many foods, it is difficult to separate the effects of these two substances. Fairweather-Tait[23] has carefully reviewed the literature. In the human, McCance et al.[24] found that sodium phytate decreased Fe absorption, Sandstead et al.[25] found no effect and Reinhold et al.[26] found that serum Fe fell initially with the feeding of phytate and then returned to normal. In the rat, Davies and Nightingale[27] and Morris and Ellis[28] observed an adverse effect of phytate on Fe absorption, while Cowan et al.[29] and Ranhotra et al.[30] found no effect. The in vitro studies of Vohra et al.[31] showed that complexes of phytate and Fe formed at pH 7.4. Sharpe et al.[32] showed that in adolescent boys, sodium phytate caused a marked reduction in Fe absorption when added to a diet containing milk.

Tsai and Lei[33] found no effect of cellulose on Fe absorption in the rat, but Ismail-Beigi et al.[34] observed a marked binding of Fe to bran fiber fractions at pH 6.8 in vitro. Widdowson and McCance[35] found that the absorption of Fe in humans was higher from white bread. In contrast, Callender and Warner[36] found that it was better from brown bread. Walker et al.[37] found no difference in absorption between bread with 70% extraction flour and that with 95 to 100% extraction flour. Bjorn-Rasmussen[38] reported that Fe absorption was reduced by 50% with the addition of bran to wheat bread. Jenkins et al.[39] found a significant fall in serum Fe with the addition of 31 g of wheat fiber per day to the diet. In rat studies, Ranhotra et al.[40] found that the addition of wheat bran to the diet reduced Fe availability. Reinhold et al.[41] reported that Fe in wholemeal bread or bran was bound to a greater extent when phytate was removed, either by acid extraction or enzymatically. This indicates that fiber largely determines the availability of divalent metals from bread.

The question of phytin or fiber having the major role in influencing the availability of Fe was addressed by Fairweather-Tait.[23] Instead of removing the phytate, as had been done in other studies, she added phytate, when needed, to equalize the phytate levels at 6.2 to 6.4 g/kg. While there was reduced Fe absorption with the phytate-containing diets in comparison to the low-phytate group, there was no difference in Fe absorption between the white, brown, and whole-meal bread groups. The greater absorption of Fe in the control group may have been due to its presence as added ferrous sulfate rather than being in the bread.

In studies with anemic pigs, Frolich and Lyso[42] fed diets containing 7% wheat bran with 60% of the Fe supplied by the bran and 40% by ferrous sulfate. The control group was fed no wheat bran and 80% of the Fe was supplied by ferrous sulfate. There was no inhibitory effect of the bran on Fe absorption at the 7% level or even at 20% in another experiment. Hemoglobin response was also studied in rats fed various components of bran by Gordon and Chao.[43] The relative availabilities of Fe from wheat bran and spinach were 124 and 53% in comparison to a control diet containing no dietary fiber. The addition of cellulose or phytic acid at levels equivalent to the amounts in wheat bran increased the relative biological value of the Fe to 126 and 124%, respectively. In ^{59}Fe uptake studies with rats, Baig et al.[44] found that 2% of dietary pectin had no effect on Fe uptake or turnover.

Fe binding to fiber in an in vitro system was studied by Leigh and Miller.[45] Fiber isolated from wheat bran had no effect on Fe availability when added to a complex meal containing beef, bread, green beans, and milk. When it was added to a semisynthetic diet, fiber, as well as ascorbate, citrate, nitrilotriacetic acid, and EDTA, effectively enhanced the availability of Fe. The other ligands in the diet appear to have a major influence on the binding of Fe in the presence of fiber. This may be one of the reasons for the wide discrepancies which exist in the studies on the effect of various substances on the availability of Fe in biological studies.

D. Amino Acids and Protein

There are marked differences in the effect of proteins from various sources on Fe availability and it is necessary to consider the origin of the protein in considering their effects.

1. Meat

Meat enhances the absorption of nonheme Fe as was first demonstrated by Layrisse et al.[46] when they added veal to a meal of vegetable origin in humans and observed improved nonheme Fe utilization. Other muscle foods have also been shown to improve the absorption of nonheme Fe, although the factor responsible has not been characterized.[47,48] Amino acids or peptides released during absorption may be involved since Martinez-Torres and Layrisse[49] found that Fe absorption from a meal of black beans was increased as much from a mixture of amino acids simulating the protein of fish as by fish protein itself. Cysteine alone, administered to humans[50] during a meal, doubled the absorption of nonheme Fe. Cook and Monsen[51] found differences in animal protein sources in improving the absorption of Fe in humans. Beef, lamb, pork, liver, fish, and chicken substitution for ovalbumin in a semisynthetic meal increased Fe absorption while milk, cheese, or egg did not.

L-Histidine, L-cysteine, and L-lysine were found to enhance the uptake of iron from ligated, in vivo segments of the duodenum of rats.[52] Derivatives of these amino acids were ineffective in influencing Fe uptake. The author suggests that the ability of these amino acids to form tridentate chelates is essential to their effectiveness in enhancing Fe uptake. There was no difference between the D- and L-forms of the amino acids, which suggests that active transport is of little significance in their effect on Fe absorption.

Kane and Miller[53] studied the effects of intact proteins and products from protein digestion on Fe availability by measuring dialyzability under simulated GI conditions. Dialyzability was high for bovine serum albumin and beef, intermediate for egg albumin, and low for soy flour, gelatin, casein, soy protein isolate, and gluten. Low molecular weight digestion products from the bovine serum albumin and beef improved Fe dialysis while preparations from casein and soy protein isolate had no effect. They proposed that the effect of proteins on Fe availability is related to the affinity of the undigested or partially digested protein for Fe and the stability of the soluble small molecular weight Fe complexes.

2. Milk and Egg Proteins

Sharpe et al.[32] observed that the addition of milk to a test diet reduced the absorption of Fe from 26.9 to 17.4%. On the other hand, Amine and Hegsted[19] reported that there was no difference in Fe absorption in rats when purified casein was increased from 5 to 40% of the diet. Cook and Monsen[51] observed no increase in Fe absorption with the substitution of milk, cheese, or egg for ovalbumin in the diet.

Milk contains three proteins which bind Fe in various degrees: lactoferrin, transferrin, and casein.[233] In human milk, lactoferrin is a major protein while transferrin concentrations are relatively low. The major Fe-binding compounds appear localized in the milk fat fraction. In cow's milk, on the other hand, lactoferrin and transferrin are low and casein is the major Fe-binding protein. The phosphate groups of casein are responsible for the Fe-binding activity. The bioavailability of Fe from human milk is in the order of 50% while it is only about 10% from cow's milk. The availability of Fe from Fe-supplemented formulas based on cow's milk may only be in the order of 3 to 4%. The difference in the availability of Fe from human milk and cow's milk may be explained by the relative differences in the proteins of the two milks, although there are still some discrepancies in the data which rule out any definite conclusion.

In their study on the effect of foods from animal sources on Fe availability, Cook and Monsen[51] (Figure 4) found egg, along with milk and cheese, to depress nonheme Fe absorption. These studies were carried out by performing multiple radioiron absorption measurements on human volunteers. The inhibiting effect of the egg yolk is not limited to the Fe in the egg, but also the availability of the Fe from all the food consumed with it is depressed. One may consider the nonheme Fe in all the food ingested in a common pool with various factors which improve or inhibit Fe availability interacting within the digestive tract.[55] It is presumably the phosphoprotein in the egg yolk, phosvitin, which binds the Fe strongly and reduces its availability. The relative biological value of Fe in spray-dried egg yolk as determined by hemoglobin regeneration in anemic rats was found to be about 85%

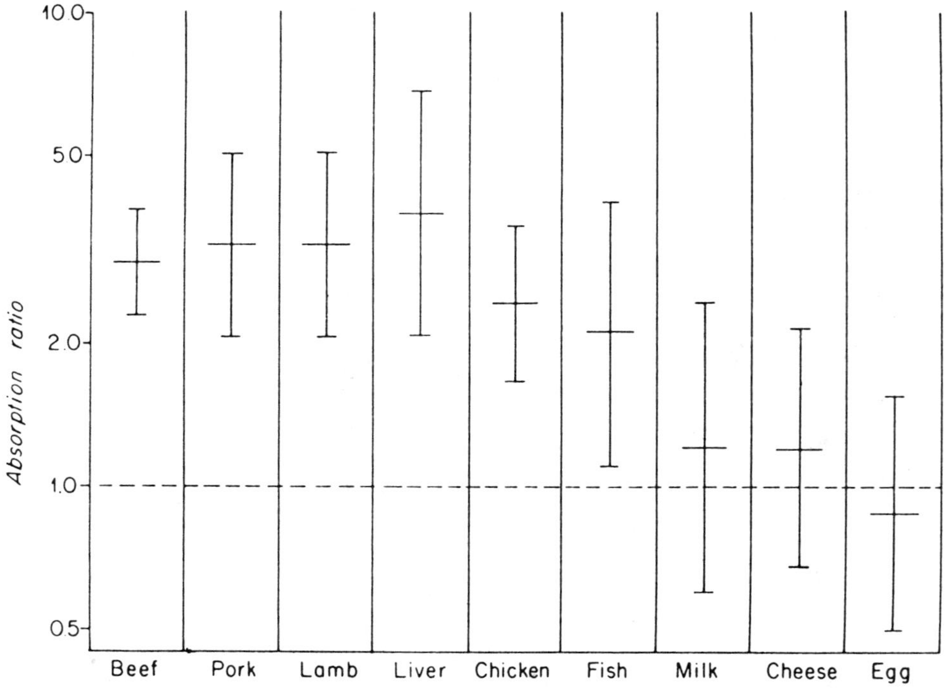

FIGURE 4. The effect on Fe absorption of substituting various animal proteins for ovalbumin contained in the SS meal. The mean absorption ratio (substituted SS/SS) + 2 SE has been plotted on a logarithmic scale for each protein substitute. (From Cook, J. D. and Monsen, E. R., *Am. J. Clin. Nutr.*, 29, 859, 1976. With permission.)

in comparison to ferrous sulfate.[56] With the addition of 0.1% ascorbic acid to the diet, there was no difference in Fe availability between egg yolk and the reference salt. In prophylactic assays with weanling rats, Miller and Nnanna[57] found relative biological values of 61% for the Fe in egg yolk when the eggs were boiled and added to dry ingredients and 64% when the yolk was pasteurized and the other diet components were steamed. When the egg yolk was baked with the other ingredients, the biological value increased to 92%. The biological value was increased, again, with the addition of ascorbic acid. It is difficult, from these studies, to categorically conclude that egg yolks definitely inhibit Fe absorption. Rossander et al.[58] studied the absorption of Fe by human subjects from typical breakfast meals. They found that the absorption decreased from 9.3 to 7.6% when an egg was served at breakfast, but the amount of Fe absorbed was practically the same due to the higher Fe content of the egg-containing breakfast.

E. Soybean Protein

In studies on the effect of chelating agents on the availability of trace elements for chicks from diets containing isolated soybean protein,[59] it was found that EDTA reduced the availability of Fe while it improved the availability of Zn and some other trace elements. It was not possible to determine whether the Fe was completely available from the soybean protein in those studies. Cook et al.[60] observed in their studies with male volunteers that the substitution of isolated soybean protein for casein and egg albumin in a semisynthetic diet reduced Fe absorption from 2.5 and 2.7% to 0.5%. The inhibitory effect was noted for a number of soybean products, such as full-fat soy flour, textured soy flour, as well as isolated soybean protein.

The question of phytic acid or soybean protein being responsible for the effect of soybean products was studied by Hallberg and Rossander.[61] Nonheme Fe absorption was reduced by 25% by the substitution of soybean flour for half of the meat protein in hamburger and even more by the use of defatted soybean flour. The removal of phytates from soy flour did not increase the absorption of nonheme Fe. The absorption of heme Fe was not influenced by the addition of soy products to the diet.

An increase in Fe absorption was noted in studies with men[62] when test meals containing either isolated soy protein or whole soybeans were baked at 200°C. The addition of meat to the diet improved Fe absorption somewhat and ascorbic acid was even more effective in improving Fe availability. In in vitro studies, Stricker et al.[63] found that the incorporation of soy protein isolate into a semisynthetic meal reduced the dialyzable Fe compared to an egg white control and that this effect was reversed by baking the soy isolate. In a number of products, the Fe availability appeared to be inversely related to the protein contribution of the soy products. In general, soy protein isolates had lower relative Fe availability than soy flours.

Thompson and Erdman[64] demonstrated that feeding a soy protein isolate can have a double effect in influencing the availability of Fe in studies with rats. Diets were fed for 13 days prior to the test meal. The presence of isolated soybean protein in the pretest diet reduced the absorption of Fe in the test period, as well as having a specific adverse effect on Fe absorption during the test period.

F. Beans

Kojima et al.[65] investigated the effect of various materials on the in vitro solubility of Fe from cooked pinto beans. Incubations were carried out at pH 2, with or without ascorbic acid, followed by an incubation at pH 6, with or without the addition of various chelating agents. The Fe could be classified into three fractions

1. That which was spontaneously soluble in the incubation procedure
2. That which can be solubilized by chelating or reducing agents
3. That which remains bound to the insoluble bean residue

These fractions represented approximately 25, 45, and 30%, respectively, of the total Fe in the beans (Figure 5). Ascorbic acid was most effective under acidic conditions. The combination of ascorbic acid and citric acid led to the solubilization of about 70% of the Fe.

G. Tannins

In a study of the absorption of Fe from breakfast meals, Rossander et al.[58] found that tea reduced the absorption of Fe to less than half the control value for a Western-type meal. Orange juice taken with the breakfast with tea more than doubled the absorption of iron from the meal. Disler et al.[66] showed the same effect previously with a rice diet. Coffee was found to decrease Fe absorption from a hamburger meal by 39% while tea caused a decrease of 64%.[67] In their in vitro study, Kojima et al.[65] found that tea severely decreased the solubility of Fe in the system. The effect of tea is presumed to be due the tannins contained in it. If this is true, tannins, which are found in many foods, might be expected to have a similar adverse effect on the availability of nonheme Fe.

H. Oxalates

In their studies with bean fractions, Kojima et al.[65] found that whole spinach suspension and insoluble spinach residue would remove from solution Fe that was previously solubilized from beans. Oxalic acid alone appeared to reduce the Fe in beans to its ferrous state and slightly decrease the amount of Fe solubilized by ascorbic acid. In feeding studies with

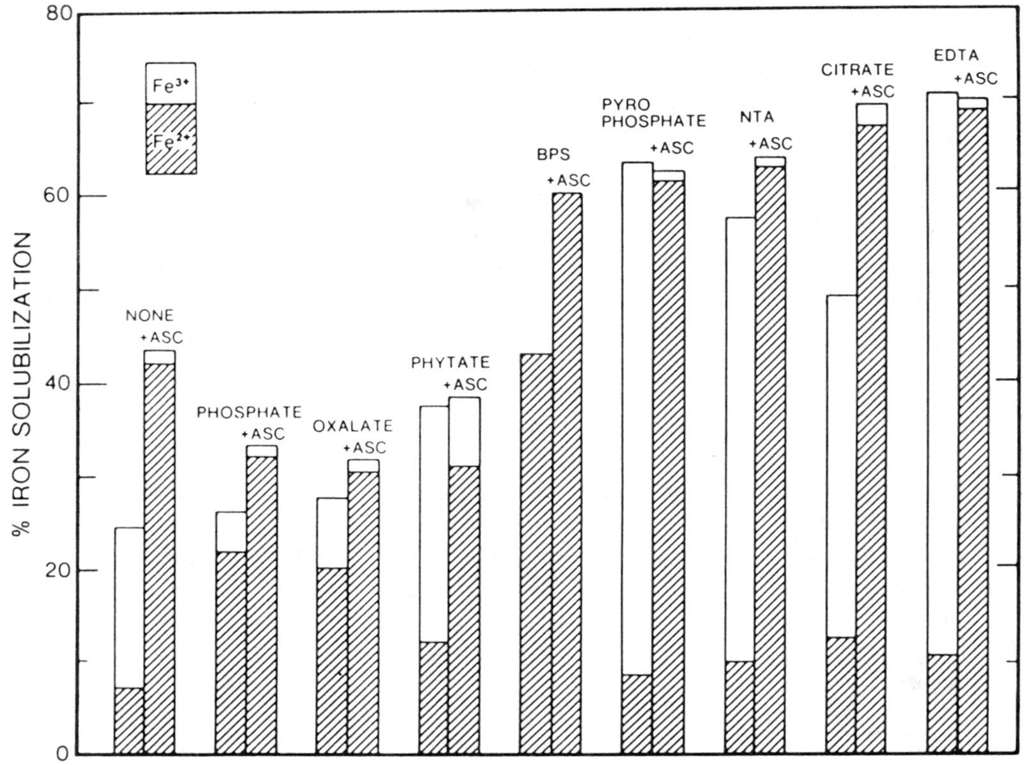

FIGURE 5. The effect of ascorbic acid, chelating agents, and combinations of ascorbic acid and chelating agents on the solubilization of Fe from bean suspension. The bean suspension was first incubated at pH 2 with (+ASC) or without ascorbic acid. The pH was then adjusted to 6 with NaOH and the chelating agents added to give a concentration of 10 m*M*. A second incubation was carried out at pH 6. The bar on the left of each pair indicates the results obtained for that chelating agent in the absence of ascorbic acid and the bar on the right (+ASC) with 10 m*M* ascorbic acid. (From Kojima, N. et al., *Am. J. Clin. Nutr.*, 34, 1392, 1981. With permission.)

anemic rats, Gordon and Chao[43] found that spinach reduced the availability of Fe to 53% of the ferrous sulfate control. Addition oxalic acid at 2.1% of the diet, or the level supplied by the spinach, improved the relative biological value to 164%, indicating that oxalic acid is not responsible for the adverse effect of spinach on Fe availability. The beneficial effect of oxalic acid alone may be due to its reducing properties.

I. Surfactants

Early work by Wissler et al.[68] indicated that polyoxyethylene sorbitan monolaurate (Tween 20) increased the GI absorption of Fe in hamsters. In double radioiron studies with human subjects, however, Brise[69] found no significant increase in Fe absorption with Tween 20, dioctylsodium sulfosuccinate, sodium lauryl sulfate, and bile acids.

J. Synthetic Chelating Agents

In experiments on the availability of trace elements from isolated soybean protein for chickens, Davis et al.[70] found that there were growth responses to the addition of EDTA to diets containing this protein and deficient in Zn, Mg or Cu, and that the response was due to improving the availability of the minerals in the soybean protein rather than being due to EDTA by itself. In the same study, there was no response in growth, hemoglobin level, or hematocrit value with the addition of EDTA to the isolated soybean protein diet while

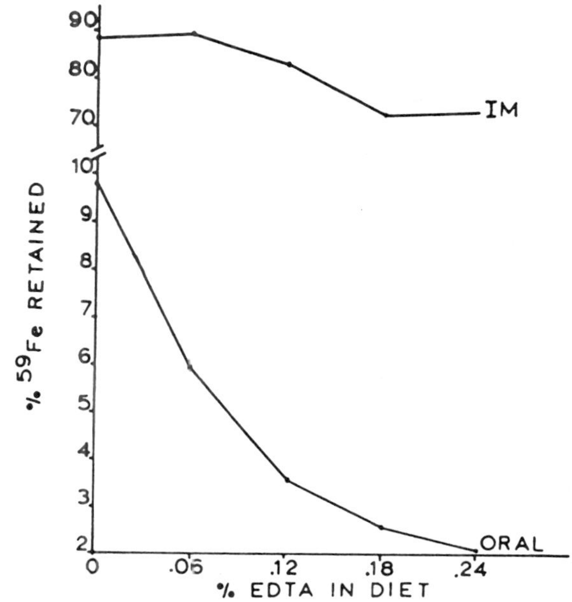

FIGURE 6. The effect of increasing levels of EDTA supplementation on ^{59}Fe absorption. (From Suso, F. A. and Edwards, H. M., *Poultry Sci.*, 47, 1417, 1968. With permission.)

Fe alone gave responses. In further studies,[59] it was found that the Fe of isolated soybean protein was not reduced by washing with water, but was reduced by about 50% by washing with a Na_2·EDTA solution. Fe from the untreated soybean protein was found to be available for growth, hemoglobin formation, and packed cell volume. In contrast, Fe which was added to the EDTA-treated soybean protein diet was not as available as that originally present. The Fe remaining in the protein after EDTA treatment and the added Fe were less available when 0.07% EDTA was added to the diet than when the diet contained no added EDTA. It appears then, that EDTA does not improve the availability of Fe, but has a tendency to reduce its availability when added to the diet. This effect was demonstrated further by Chio et al.[71] when they were able to induce a deficiency of Fe in chickens by adding 1 or 1.5% of Na_2·EDTA to an adequate practical ration. A dystrophic strain of New Hampshire chicks to which the EDTA was fed developed white feathers and showed reduced hematocrit and hemoglobin values. These abnormalities were corrected by adding Fe to the diet.

The absorption of ^{59}Fe was studied by Suso and Edwards[72] with a whole body counting technique after the administration of the isotope to chickens which had been fed various levels of chelating agents. The retention of Fe was decreased from approximately 10% without a chelating agent to about 2% with the addition of 0.24% EDTA to the diet, as shown in Figure 6. Fe absorption was also reduced by HEDTA, DTPA, and EDDHA, but it was slightly improved by the addition of HEIDA, which has a lower stability constant for Fe than the other compounds. None of the chelates tested influenced the biological half-life.

In three experiments with poults,[73] various chelates were fed in a Zn-deficient soy protein diet. EDTA corrected the Zn deficiency as shown by improved growth and reduced perosis score, but the packed cell volume was markedly decreased. NTA, citric acid, and ascorbic acid were ineffective in counteracting this effect.

Larsen et al.[74] fed Na_2·EDTA at a level of 0.1% to young rats and studied the absorption, metabolism, and excretion of radioactive ferric chloride given by stomach tube. The presence

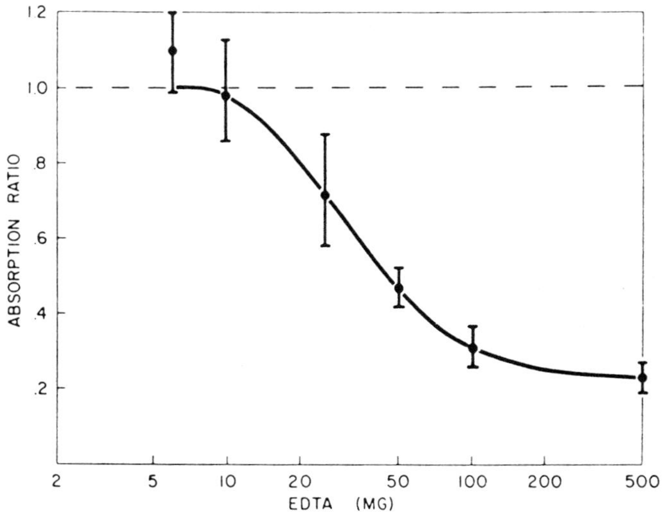

FIGURE 7. Effect of EDTA on Fe absorption from the STD meal. The ratio of Fe absorption with and without EDTA is plotted against the amount added. The vertical bars represent the limits of ± 1 SE. (From Cook, J. D. and Monsen, E. R., *Am. J. Clin. Nutr.*, 30, 1166, 1977. With permission.)

of EDTA decreased the absorption of Fe and increased its excretion in the urine. When EDTA was not fed until after the ^{59}Fe had been absorbed, the Fe appeared to be increased in the blood and spleen, and to be removed from the liver. Brise and Hallberg[7] also found that EDTA depressed Fe absorption in human subjects. By the use of whole body counting after the administration of ^{59}Fe to humans, Davis and Deller[75] also studied Fe absorption in humans. EDTA and DTPA had no significant effect on Fe absorption when given in a 1:1 ratio, but reduced it when given in a ratio of 10:1 or 50:1. Fructose, on the other hand, had no effect at a 1:1 concentration but increased the absorption at similar higher ratios.

Cook and Monsen[76] studied Fe absorption in normal men by measuring the incorporation of ^{55}Fe and ^{59}Fe in red cells in humans given both standard meals and semisynthetic diets. The addition of 50 mg EDTA to test meals containing 4.1 mg Fe reduced absorption by approximately one half from both the standard and semisynthetic diets. Exchange studies with the two isotopes showed a complete exchange of ferric EDTA with nonheme Fe in the diet. As shown in Figure 7, the absorption ratio decreased from 1.0 at 10 mg of EDTA to approximately 0.2 with about 100 mg of EDTA. The authors pointed out that they were not measuring absorption as such, but from a nutritional standpoint, whole body retention of Fe is the important functional measurement.

While high levels of EDTA unquestionably reduce the availability of Fe, lower levels have been shown to have no effect on absorption. Fritz et al.,[8] in studying the availability of Fe from various sources, reported EDTA, dihydrogen ferrous salt to have a biological value of 97 to 100 in studies with chicks and rats. Layrisse and Martinez-Torres[77] found that the Fe of Fe(III)·EDTA exchanged completely with the vegetable food Fe in the lumen of the gut but that the absorption of both intrinsic and extrinsic food Fe is higher than expected from other Fe salts. The absorption from the EDTA complex was approximately twice that from ferrous sulfate and they suggested that in the order of 10 to 15 mg of Fe per day in the complex form would be an adequate human supplement. Viteri et al.[78] also reported that Fe absorption from NaFeEDTA was better than from the reference ferric sulfate when 5 mg of Fe from various sources was compared with that in both adults and children. Their work showed that NaFe·EDTA, $Fe_2(SO_4)_3$, and nonheme food Fe form a common

Table 3
TISSUE DISTRIBUTION 24 HR AFTER I.P. INJECTION, 6 Mg Fe/kg IN NORMAL RATS (% OF DOSE)

Ferric compound	No. of animals	Blood	Urine	Feces	Gut	Carcass	Marrow	Liver	Cumulative
Citrate	5	3.2	6	0.7	8	81	ND	ND	99
NTA	3	3.2	9	0.3	10	75	ND	ND	98
NEIDA	4	4.4	5	0.3	14	85	ND	ND	107
CDTA	5	0.8	90	6	3	3	1	0.3	105
DTPA	6	1.5	83	4	3	3	ND	ND	92
ECTA	6	0.5	83	3	4	ND	ND	ND	91
CHCEDG	5	1.5	58	18	7	10	4	4	101
EHCHG	5	3.9	58	15	14	6	4	2	103
EHPG	10	3.1	54	12	16	1	7	2	95
CH_3EHPG	4	1.5	33	29	19	ND	ND	ND	83
ClEHPG	5	3.7	37	19	12	14	ND	ND	86

Note: ND = not determined.

From Rubin, M. et al., *Biochem. Med.*, 3, 271, 1970. With permission.

pool which is different from the heme pool, but which is made more available by the presence of NaFe·EDTA. In studies on women in India, MacPhail et al.[79] reported that Fe absorption from maize porridge fortified with Fe(III)·EDTA was more than twice that from porridge fortified with $FeSO_4 \cdot 7\ H_2O$. Martínez-Torres et al.[80] found that the absorption of Fe from ferrous sulfate might vary from 2 to 30% depending upon the food in which it was studied, while the absorption from Fe(III)·EDTA remained practically constant.

Ferripolyphosphate, an aqueous complex of ferric chloride and sodium polyphosphate, precipitates the proteins from acid whey[81] at pH 4.6. Both aqueous ferripolyphosphate and the whey protein complex were found to have 92 to 100% Fe availability in comparison to ferrous sulfate, although the ferripolyphosphate in gel form was only 50 to 60% effective. When used at high levels, rats developed less severe lesions with ferripolyphosphate than with equivalent levels of Fe from ferrous sulfate. The Fe in the polyphosphate complexes was so tightly bound that there was no off-flavor produced in milk products when 10 mg of Fe was added to milk in the polyphosphate-whey protein complex and held for 2 weeks.

Studies by Bates et al.[82,83] of the kinetics and mechanism of Fe exchange between chelates and human transferrin relates to the mechanism of action of chelates in Fe utilization. While equilibrium concentrations are related to the magnitude of the relative stability constants, the kinetics of transfer appear to be regulated by stearic factors. The times required for equimolar concentrations of three typical chelates to half-saturate the sites is nitrilotriacetate, 3 sec; citrate, 8 hr; EDTA, about 4 days. They proposed a reaction sequence involving a ternary complex of chelate Fe-transferrin as an intermediate. The rate of removal of Fe from ferric transferrin was in the order: EDTA > citrate > nitrilotriacetate. The latter two chelates reached equilibrium more rapidly.

Rubin et al.[84] studied the metabolism of parentally administered Fe chelates in the rat. Fe complexed with citrate, NTA, or NEIDA was largely distributed in the carcass after 24 hr while Fe with the other chelates was largely excreted in the urine (Table 3). The strong anionic hydrophilic Fe chelates provided the least Fe cleavage and the highest urinary excretion. A second group of less polar Fe chelates of high stability followed a distribution pattern which provided GI as well as urinary Fe excretion. Fe from a third group of Fe chelates of binding strength lower than that of transferrin, the specific Fe-binding protein

Table 4
DISTRIBUTION OF ORAL Fe AFTER 2 DAYS, PERCENTAGE OF DOSE, NORMAL RATS (6 mg Fe/kg)

Ferric compound	Feces and gut	Blood	Urine	No. of animals
Chloride	76	5.0	0.5	7
Citrate	72	2.5	0.3	3
Fructose	74	2.8	0.1	3
NTA	69	3.4	0.7	3
HEIDA	81	0.5	0.2	3
DHEG	76	2.7	0	3
EDTA	72	0.6	3.6	3
CDTA	70	4.5	9.6	3
DTPA	75	1.6	6.2	6
CHCEDG	68	4.5	2.0	3
EHPG	82	1.0	4.0	3
CH_3EHPG	69	0.1	3.0	3
ClEHPG	71	4.7	3.0	4

From Princiotto, J. V. et al., *Biochem. Med.*, 3, 289, 1970. With permission.

of the plasma, was retained and widely distributed after parenteral injection. The anemic animal utilized Fe more efficiently than the normal animal.

Princiotto et al.[85] extended their study of chelate metabolism to the absorption of oral chelated Fe (Table 4). The unabsorbed Fe in the gut and feces 48 hr after the administration varied from 68 to 82% of the dose, although the individual variability probably prevented these differences from being significant. All of the compounds bind Fe tightly so that it is retained by the carrier in the presence of transferrin. However, the compounds differ in polarity of the complexes. The net charge for DTPA is binegative, compared with a single negative charge for EDTA, HOEDTA, and CDTA, and the neutral structure of the Fe chelates of the EHPG series. The authors observed that the best absorption occurred in the undissociated state. Within a homologous group it may be anticipated that features which confer increased lipid solubility will enhance membrane permeability. The sequence EDTA, HOEDTA, and CDTA shows a gradual shift toward a more lipotropic character and an increase in the absorption of the intact Fe chelate.

Neither the anionic Fe chelates of the EDTA type or the neutral compounds of the EHPG series show any advantage in Fe availability in comparison to ferric chloride.

However, NTA, HEIDA, and DHEG were as effective as ferric chloride in providing Fe to the anemic animal. This has been confirmed in other studies and indicates that the Fe from these three chelating agents is highly available at the site at which it is needed for hemoglobin synthesis (Table 5).

While our discussion thus far has been mainly concerned with the nutritional availability of Fe, there is also concern about the use of chelates for the alleviation of an Fe overload as might result from multiple blood transfusions. Hershko et al.[86] screened some 15 compounds with a high affinity to ferric Fe for in vivo Fe-chelating efficiency in the rat. Of the several compounds tested (Table 6), ethylenediamine-*N,N'*-*bis*(*o*-hydroxyphenylglycine) (EHGP) was one of the most potent. A high selective affinity of a compound for ferric Fe is a basic requirement in the selection of Fe-chelating drugs. Thermodynamic constants do not permit prediction of the bioavailability, toxicity, and metabolism of Fe-chelating drugs. They cite as factors critical for determining in vivo behavior of Fe-chelating agents as their

Table 5
DISTRIBUTION OF ORAL IRON AFTER 2 DAYS, PERCENTAGE OF DOSE, ANEMIC ANIMALS (6 mg Fe/kg)

Ferric compound	Feces and gut	Blood	Urine	No. of animals
Chloride	67	27	0.6	11
Citrate	69	29	1.8	5
Fructose	71	14	1.2	3
NTA	68	25	0.2	6
HEIDA	57	32	0.4	9
DHEG	64	33	0.5	3
EDTA	66	14	6.5	4
CDTA	73	17	7.3	3
DTPA	67	13	2.0	10
EHPG	63	15	9.3	8
CH_3EHPG	83	9	5.3	3
ClEHPG	70	14	5.9	4

From Princiotto, J. V. et al., *Biochem. Med.*, 3, 289, 1970. With permission.

Table 6
EXCRETION AND ORGAN DISTRIBUTION OF RADIOIRON-DRUG COMPLEX (% OF TOTAL)

Compound	Urine	Stool	Blood	Liver	Residual
Salicylhydroxamic acid	0.3 ± 0	3.5 ± 0.2	31.8 ± 0.5	11.5 ± 1.2	52.9 ± 3.8
8-Hydroxyquinoline	0.1 ± 0	3.6 ± 0.2	24.6 ± 3.2	16.7 ± 1.5	55.0 ± 5.7
Pyrocatechol violet	0.4 ± 0	3.7 ± 0	22.6 ± 0.5	11.6 ± 0.5	61.7 ± 0.7
4,5-Benzotropolone	1.9 ± 0.7	4.9 ± 1.9	37.3 ± 2.3	13.7 ± 0.7	42.2 ± 1.9
2,3-Dihydroxynaphthalene	1.6 ± 0.1	6.0 ± 0.1	25.3 ± 0.9	9.1 ± 1.5	58.0 ± 1.0
2-Piperidinomethyl-3-hydroxy-6-hydroxymethyl-4H-pyran-4-one	4.4 ± 0.1	10.7 ± 0.3	35.1 ± 2.5	11.2 ± 2.1	38.6 ± 1.9
2,3-Dihydroxybenzoic acid	5.1 ± 0.1	9.4 ± 0.2	21.7 ± 0.2	8.9 ± 2.9	54.9 ± 1.3
Cholylhydroxamic acid	8.1 ± 0.2	11.2 ± 0.3	27.9 ± 0.1	8.3 ± 0.7	44.4 ± 1.2
N^1,N^8-*bis*(2,3-dihydroxybenzoyl) spermidine · HCl	7.2 ± 0.1	25.2 ± 0.5	1.7 ± 0.1	2.4 ± 0.1	63.5 ± 1.4
N,N'-*bis*(2,3-dihydroxybenzoyl)-1,6-diaminohexane	17.3 ± 1.8	47.3 ± 4.9	8.0 ± 1.2	6.5 ± 0.1	20.9 ± 6.1
Pyridoxyl isonicotinoyl hydrazone	4.3 ± 0.4	32.8 ± 5.9	32.0 ± 3.1	9.6 ± 1.3	20.3 ± 11.2
Ethylenediamine-*N,N'*-*bis*(o-hydroxy-phenylglycine) (EHPG)	41.7 ± 0.8	22.2 ± 1.4	15.3 ± 2.0	7.0 ± 1.1	13.8 ± 1.6
Ethylenediamine-*N,N'*-*bis*(2-hydroxy-5-carboxyphenylglycine	6.0 ± 1.0	60.1 ± 10.3	8.3 ± 1.3	3.3 ± 0.3	22.3 ± 4.3
N-Acetyldesferrioxamine	58.9 ± 2.0	16.2 ± 0.6	12.8 ± 1.4	4.5 ± 0.2	13.4 ± 0.9
Picolinaldehyde Thiosemicarbazone	56.1 ± 6.1	29.9 ± 2.1	2.5 ± 0.4	7.9 ± 2.9	4.1 ± 1.7
Desferrioxamine (DF)	69.5 ± 7.6	5.5 ± 0.4	7.9 ± 1.2	3.3 ± 1.3	13.7 ± 5.9
Control	0.1 ± 0	4.8 ± 0.1	27.1 ± 3.6	12.4 ± 1.7	54.6 ± 2.6

From Hershko, C. et al., *Br. J. Haematol.*, 51, 251, 1982. With permission.

solubility, membrane permeability, and ability to compete with other natural Fe-chelating systems such as transferrin.

REFERENCES

1. **McGilvery, R. W.,** *Biochemistry, A Functional Approach,* W. B. Saunders, Philadelphia, 1970.
2. **Hallberg, L.,** Bioavailability of dietary iron in man, in *Annual Review of Nutrition,* Vol. 1, Darby, W. J., Broquist, H. P., and Olson, R. E., Eds., Annual Reviews, Palo Alto, Calif., 1981, 123.
3. **Morris, E. R.,** An overview of current information on bioavailability of dietary iron to humans, *Fed. Am. Soc. Exp. Biol. Fed. Proc.,* 42, 1716, 1983.
4. **Finch, C. A. and Cook, J. D.,** Iron deficiency, *Am. J. Clin. Nutr.,* 39, 471, 1984.
5. **Hallberg, L., Bjorn-Rasmussen, E., Howard, L., and Rossander, L.,** Dietary heme iron absorption. A discussion of possible mechanisms for the absorption-promoting effect of meat and for the regulation of iron absorption, *Scand. J. Gastroenterol.,* 14, 769, 1979.
6. **Hallberg, L. and Rossander, L.,** Improvement in iron nutrition in developing countries: comparison of adding meat, soy protein, ascorbic acid, citric acid, and ferrous sulphate on iron absorption from a simple Latin American-type of meal, *Am. J. Clin. Nutr.,* 39, 577, 1984.
7. **Brise, H. and Hallberg, L.,** Absorbability of different iron compounds, *Acta Med. Scand. Suppl.,* 376, 23, 1962.
8. **Fritz, J. C., Pla, G. W., Roberts, T., Boehne, J. W., and Hove, E. L.,** Biological availability in animals of iron from common dietary sources, *J. Agric. Food Chem.,* 18, 647, 1970.
9. **Van Campen, D.,** Effect of histidine and ascorbic acid on the absorption and retention of ^{59}Fe by iron-depleted rats, *J. Nutr.,* 102, 165, 1972.
10. **Gipp, W. F., Pond, W. G., Kallfelz, F. A., Tasker, J. B., Van Campen, D. R., Krook, L., and Visek, W. J.,** Effect of dietary copper, iron and ascorbic acid levels on hematology, blood and tissue copper, iron and zinc concentrations and ^{64}Cu and ^{59}Fe metabolism in young pigs, *J. Nutr.,* 104, 532, 1974.
11. **Hill, C. H. and Starcher, B.,** Effect of reducing agents on copper deficiency in the chick, *J. Nutr.,* 85, 271, 1965.
12. **Miski, A. M. and Kratzer, F. H.,** Effect of dietary ascorbic acid on iron utilization in the growing chick, *Poult. Sci.,* 55, 454, 1976.
13. **Brise, H. and Hallberg, L.,** Effect of ascorbic acid on iron absorption, *Acta Med. Scand. Suppl.,* 376, 51, 1962.
14. **Lee, P. C., Ledwich, J. R., and Smith, D. C.,** Large and small doses of ascorbic acid in the absorption of ferrous iron, *Can. Med. Assoc. J.,* 97, 181, 1967.
15. **Grebe, G., Martinez-Torres, C., and Layrisse, M.,** Effect of meals and ascorbic acid on the absorption of a therapeutic dose of iron as ferrous and ferric salts, *Curr. Ther. Res.,* 17, 382, 1975.
16. **Hungerford, D. M., Jr. and Linder, M. C.,** Interactions of pH and ascorbate in intestinal iron absorption, *J. Nutr.,* 113, 2615, 1983.
17. **Milne, D. B. and Omaye, S. T.,** Effects of vitamin C on copper and iron metabolism in the guinea pig, *Int. J. Vitam. Nutr. Res.,* 50, 301, 1980.
18. **Brise, H. and Hallberg, L.,** Effect of succinic acid on iron absorption, *Acta Med. Scand. Suppl.,* 376, 59, 1962.
19. **Amine, E. K. and Hegsted, D. M.,** Effect of diet on iron absorption in iron-deficient rats, *J. Nutr.,* 101, 927, 1971.
20. **Amine, E. K. and Hegsted, D. M.,** The effect of dietary carbohydrates and fats on inorganic iron absorption, *J. Agric. Food Chem.,* 23, 204, 1975.
21. **Christopher, J. P., Hegenauer, J. C., and Saltman, P. D.,** Iron metabolism as a function of chelation, in *Trace Element Metabolism in Animals, Vol. 2,* Proc. 2nd Int. Symp. Trace Element Metabol. Anim., Madison, Wisc., 1973, 133.
22. **Bates, G. W., Boyer, J., Hegenauer, J. C., and Saltman, P.,** Facilitation of iron absorption by ferric fructose, *Am. J. Clin. Nutr.,* 25, 983, 1972.
23. **Fairweather-Tait, S. J.,** The effect of different levels of wheat bran on iron absorption in rats from bread containing similar amounts of phytate, *Br. J. Nutr.,* 47, 243, 1982.
24. **McCance, R. A., Edgecombe, C. N., and Widdowson, E. M.,** Phytic acid and iron absorption, *Lancet,* ii, 126, 1943.
25. **Sandstead, H. H., Munoz, J. M., Jacob, R. A., Klevay, L. M., Reck, S. J., Logan, G. M., Jr., Dintzis, F. R., Inglett, G. E., and Shuey, W. C.,** *Am. J. Clin. Nutr.,* 31, 5180, 1978.
26. **Rheinhold, J. G., Nasr, K., Lahimgarzadeh, A., and Hedayati, H.,** Effects of purified phytate and phytate-rich bread upon metabolism of zinc, calcium, phosphorus and nitrogen in man, *Lancet,* 1, 283, 1973.
27. **Davies, N. T. and Nightingale, R.,** The effect of phytate on intestinal absorption and secretion of zinc, and whole-body retention of Zn, copper, iron and manganese, *Br. J. Nutr.,* 34, 243, 1975.
28. **Morris, E. R. and Ellis, R.,** Bioavailability to rats of iron and zinc in wheat bran: response to low-phytate bran and effect of the phytate/zinc molar ratio, *J. Nutr.,* 110, 2000, 1980.

29. **Cowan, J. W., Esfahani, M., Salji, J. P., and Azzam, S. A.,** Effect of phytate on iron absorption in the rat, *J. Nutr.*, 90, 423, 1966.
30. **Ranhotra, G. S., Loewe, R. J., and Puyat, L. V.,** Effect of dietary phytic acid on the availability of iron and phosphorus, *Cereal Chem.*, 51, 323, 1974.
31. **Vohra, P., Gray, G. A., and Kratzer, F. H.,** Phytic acid-metal complexes, *Proc. Soc. Exp. Biol. Med.*, 120, 447, 1965.
32. **Sharpe, L. M., Peacock, W. C., Cooke, R., and Harris, R. S.,** The effect of phytate and other food factors on iron absorption *J. Nutr.*, 41, 433, 1950.
33. **Tsai, R. C. Y. and Lei, K. Y.,** Dietary cellulose, zinc and copper: effects on tissue levels of trace minerals in the rat, *J. Nutr.*, 109, 1117, 1979.
34. **Ismail-Beigi, F., Faradjid, B., and Reinhold, J. G.,** Binding of zinc and iron to wheat breads, wheat bran and their components, *Am. J. Clin. Nutr.*, 30, 1721, 1977.
35. **Widdowson, E. M. and McCance, R. A.,** Iron exchanges of adults on white- and brown-bread diets, *Lancet*, i, 588, 1942.
36. **Callender, S. T. and Warner, G. T.,** Iron absorption from brown bread, *Lancet*, i, 546, 1970.
37. **Walker, A. R. P., Fox, F. W., and Irving, J. T.,** Studies in human mineral metabolism. I. The effect of bread rich in phytate phosphorus on the metabolism of certain mineral salts with special reference to calcium, *Biochem. J.*, 42, 452, 1948.
38. **Bjorn-Rasmussen, E.,** Iron absorption from wheat bread, influence of various amounts of bran, *Nutr. Metab.*, 16, 101, 1974.
39. **Jenkins, D. J. A., Hill, M. S., and Cummings, J. H.,** Effect of wheat fiber on blood lipids, fecal steroid excretion and serum iron, *Am. J. Clin. Nutr.*, 28, 1408, 1975.
40. **Ranhotra, G. S., Lee, C. and Gelroth, J. A.,** Bioavailability of iron in some commercial variety breads, *Nutr. Rep. Int.*, 19, 851, 1979.
41. **Reinhold, J. G., Ismail-Beigi, F., and Faradji, B.,** Fibre vs phytate as determinant of the availability of calcium, zinc and iron of breadstuffs *Nutr. Rep. Int.*, 12, 75, 1975.
42. **Frolich, W. and Lyso, A.,** Bioavailability of iron from wheat bran in pigs, *Am. J. Clin. Nutr.*, 37, 31, 1983.
43. **Gordon, D. T. and Chao, L. S.,** Relationship of components in wheat bran and spinach to iron bioavailability in the anemic rat, *J. Nutr.* 114, 526, 1984.
44. **Baig, M. M., Burgin, C. W., and Cerda, J. J.,** Effect of dietary pectin on iron absorption and turnover in the rat, *J. Nutr.*, 113, 2385, 1983.
45. **Leigh, M. J. and Miller, D. D.,** Effects of pH and chelating agents on iron binding by dietary fiber: implications for iron availability, *Am. J. Clin. Nutr.*, 38, 202, 1983.
46. **Layrisse, M., Martinez-Torres, C., and Roche, M.,** Effect of interaction of various foods on iron absorption, *Am. J. Clin. Nutr.*, 21, 1175, 1968.
47. **Hallberg, L. and Rossander, L.,** Improvement in iron nutrition in developing countries: comparison of adding meat, soy protein, ascorbic acid, citric acid, and ferrous sulphate on iron absorption from a simple Latin American-type of meal, *Am. J. Clin. Nutr.*, 39, 577, 1984.
48. **Bjorn-Rasmussen, E. and Hallberg, L.,** Effect of animal proteins on the absorption of food iron in man, *Nutr. Metab.*, 23, 192, 1979.
49. **Martinez-Torres, C. and Layrisse, M.,** Effect of amino acids on iron absorption from a staple vegetable food, *Blood*, 35, 669, 1970.
50. **Martinez-Torres, C., Romans, E., and Layrisse, M.,** Effect of cysteine on iron absorption in man, *Am. J. Clin. Nutr.*, 34, 322, 1981.
51. **Cook, J. D. and Monsen, E. R.,** Food iron absorption in human subjects. III. Comparison of the effect of animal proteins on nonheme iron absorption, *Am. J. Clin. Nutr.*, 29, 859, 1976.
52. **Van Campen, D.,** Enhancement of iron absorption from ligated segments of rat intestine by histidine, cysteine, and lysine: effect of removing ionizing groups and of stereoisomerism, *J. Nutr.*, 103, 139, 1973.
53. **Kane, A. P. and Miller, D. D.,** In vitro estimation of the effects of selected proteins on iron bioavailability, *Am. J. Clin. Nutr.*, 39, 393, 1984.
54. **Lonnerdal, B., Keen, C. L., and Hurley, L. S.,** Iron, copper, and manganese in milk, in *Annual Review of Nutrition*, Vol. 1, Darby, W. J., Broquist, H. P., and Olson, R. E., Eds., Annual Reviews, Palo Alto, Calif., 1981, 149.
55. **Hallberg, L.,** The pool concept in food iron absorption and some of its implications, *Proc. Nutr. Soc.*, 33, 285, 1974.
56. **Miller, J. and McNeal, L. S.,** Bioavailability of egg yolk iron measured by hemoglobin regeneration in anemic rats, *J. Nutr.*, 113, 115, 1983
57. **Miller, J. and Nnanna, I.,** Bioavailability of iron in cooked egg yolk for maintenance of hemoglobin levels in growing rats, *J. Nutr.*, 113, 1169, 1983.
58. **Rossander, L., Hallberg, L., and Bjorn-Rasmussen, E.,** Absorption of iron from breakfast meals, *Am. J. Clin. Nutr.*, 32, 2484, 1979.

59. **Davis, P. N., Norris, L. C., and Kratzer, F. H.,** Iron deficiency studies in chicks using treated isolated soybean protein diets, *J. Nutr.*, 78, 445, 1962.
60. **Cook, J. D., Morck, T. A., and Lynch, S. R.,** The inhibitory effect of soy products on nonheme iron absorption in man. *Am. J. Clin. Nutr.*, 34, 2622, 1981.
61. **Hallberg, L. and Rossander, L.,** Effect of soy protein on nonheme iron absorption in man, *Am. J. Clin. Nutr.*, 36, 514, 1982.
62. **Cook, J. D., Morck, T. A., and Lynch, S. R.,** The inhibitory effect of soy products on non-heme iron absorption in man, *Am. J. Clin. Nutr.*, 36, 219, 1982.
63. **Stricker, B. R., Miller, D. D., and Van Campen, D.,** In vitro estimation of iron availability in meals containing soy products, *J. Nutr.*, 112, 1696, 1982.
64. **Thompson, D. B. and Erdman, J. W., Jr.,** The effect of soy protein isolate in the diet on retention by the rat of iron from radiolabeled test meals, *J. Nutr.*, 114, 307, 1984.
65. **Kojima, N., Wallace, D., and Bates, G. W.,** The effect of chemical agents, beverages, and spinach on the in vitro solubilization of iron from cooked pinto beans, *Am. J. Clin. Nutr.*, 34, 1392, 1981.
66. **Disler, P. B., Lynch, S. R., Charlton, R. W., Torrence, J. D., Bothwell, T. H., Walker, R. B., and Mayet, F.,** The effect of tea on iron absorption, *Gut,* 16, 193, 1975.
67. **Morck, T. A., Lynch, S. R., and Cook, J. D.,** Inhibition of food iron absorption by coffee, *Am. J. Clin. Nutr.*, 37, 416, 1983.
68. **Wissler, R. W., Bethard, W. F., Barker, P., and Mori, H. D.,** Effects of polyoxyethylene sorbitan monolaurate (Tween 20) upon gastrointestinal iron absorption in hamsters, *Proc. Soc. Exp. Biol. Med.*, 86, 170, 1954.
69. **Brise, H.,** Influence of meals on iron absorption in oral iron therapy, *Acta Med. Scand. Suppl.*, 376, 39, 1962.
70. **Davis, P. N., Norris, L. C., and Kratzer, F. H.,** Interference of soy protein with the utilization of trace minerals, *J. Nutr.*, 77, 217, 1962.
71. **Chio, L. F., Bunden, K., Vohra, P., and Kratzer, F. H.,** An abnormal requirement for iron in dystrophic chickens, *Poult. Sci.*, 55, 808, 1976.
72. **Suso, F. A. and Edwards, H. M.,** Influence of various chelating agents on absorption of ^{60}Co, ^{59}Fe, ^{54}Mn and ^{65}Zn by chickens, *Poult. Sci.*, 47, 1417, 1968.
73. **Kratzer, F. H.,** Zinc, iron, and chelate interaction in poults and chicks, in *Trace Element Metabolism in Animals,* Vol. 2, Proc. 2nd Int. Symp. Trace Element Metab. Anim., Madison, Wisc., 1973, 527.
74. **Larsen, B. A., Bidwell, R. G. S., and Hawkins, W. W.,** The effect of ingestion of disodium ethylenediamine-tetraacetate on the absorption and metabolism of radioactive iron by the rat, *Can. J. Biochem. Physiol.*, 38, 51, 1960.
75. **Davis, P. S. and Deller, D. J.,** Effect of orally administered chelating agents EDTA, DTPA, and fructose on radioiron absorption in man, *Aust. Ann. Med.*, 16, 70, 1967.
76. **Cook, J. D. and Monsen, E. R.,** Food iron absorption in man. II. The effect of EDTA on absorption of non-heme iron, *Am. J. Clin. Nutr.*, 29, 614, 1976.
77. **Layrisse, M. and Martinez-Torres, C.,** Fe(III)-EDTA complex as iron fortification, *Am. J. Clin. Nutr.*, 30, 1166, 1977.
78. **Viteri, F. E., Garcia-Ibanez, R., and Torun, R.,** Sodium iron NaFeEDTA as an iron fortification compound in Central America. Absorption studies, *Am. J. Clin. Nutr.*, 31, 961, 1978.
79. **MacPhail, A. P., Bothwell, T. H., Torrance, J. D., Derman, D. P., Bezwoda, W. R., Charlton, R. W., and Mayet, F.,** Factors affecting the absorption of iron from Fe(III)EDTA, *Br. J. Nutr.*, 45, 215, 1981.
80. **Martinez-Torres, C., Ramano, E. L., Renzi, M., and Layrisse, M.,** Fe(III)EDTA complex as iron fortification. Further studies, *Am. J. Clin. Nutr.*, 32, 809, 1979.
81. **Jones S. B., Kalan, E. B., Jones, T. C., Hazel, J. F., Edmondson, L. R., Booth, A. N., and Fritz, J. C.,** Ferripolyphosphate-whey protein powders. Their potential as nutritional iron supplements, *J. Agric. Food Chem.*, 23, 981, 1975.
82. **Bates, G. W., Billups, C., and Saltman, P.,** The kinetics and mechanism of iron (III) exchange between chelates and transferrin. I. The complexes of citrate and nitrilotriacetic acid, *J. Biol. Chem.*, 242, 2810, 1967.
83. **Bates G. W., Billups, C., and Saltman, P.,** The kinetics and mechanism of iron(III) exchange between chelates and transferrin. II. The presentation and removal with ethylenediaminetetracetate, *J. Biol. Chem.*, 242, 2816, 1967.
84. **Rubin, M., Pachtman, E., Aldridge, M., Zapolski, E. J., Bagley, D. H., Jr., and Princiotto, J. V.,** The metabolism of parenteral iron chelates, *Biochem. Med.*, 3, 271, 1970.
85. **Princiotto, J. V., Zapolski, E. J., Bagley, D. H., Jr., Laskey, A., Morgan, R., and Rubin, M.,** Absorption of oral chelated iron, *Biochem. Med.*, 3, 289, 1970.
86. **Hershko, C., Grady, R. W., and Link, G.,** Evaluation of iron-chelating agents in an *in vivo* system: potential usefulness of EHPG, a powerful iron-chelating drug, *Br. J. Haematol.*, 51, 251, 1982.

Chapter 8

Copper

I. INTRODUCTION

Cu is absorbed from the duodenum in man and chicks, from the upper jejunum in dogs, from the small intestine and colon in pigs, and from the stomach and small intestine in rats.[1] The mechanism of Cu absorption has not been elucidated. Starcher[2] described a protein of about 10,000 mol wt which has Cu-binding ability in the duodenum of the chick, and it may be involved in the absorption process. Cu absorption is inhibited by other ions, probably at the level of the intestine. After absorption, Cu is found in the plasma bound with a protein, ceruloplasmin, in most species other than chicks and turkeys. Ceruloplasmin has about a 151,000 mol wt and binds eight atoms of Cu per molecule.

The absorption and retention of Cu is influenced greatly by the presence of other competing ions in the gastrointestinal tract and also by a number of organic substances. The nature and action of these substances will be discussed.

II. PHYTIC ACID

In a study of the effect of soybean protein on the utilization of trace elements, Davis et al.[3] observed slight reductions in growth and hemoglobin levels in chicks fed a diet containing isolated soybean protein with a marginal level of Cu. The growth and hemoglobin levels were improved with either added Cu or EDTA. This was interpreted to indicate that the Cu in the isolated soybean protein was poorly available and the addition of EDTA improved Cu availability as it did with Zn. Lo et al.[4] reported that Cu was equally available from isolated soybean protein and Cu carbonate for the repletion of serum and liver Cu levels in rats depleted of Cu by feeding them a Cu-deficient diet for 28 days. In in vitro studies, Davis et al.[5] found that the Cu content of isolated soybean protein could be lowered by repeated washing with EDTA, but not to the same extent as Zn or Ca. If one assumes that phytic acid in isolated soybean protein is responsible for the majority of the Cu-binding activity, it is probable that its effect is minimal on Cu availability. This was also shown by the work of Lyon,[6] who found that Cu did not readily appear to form a salt with phytate and remained in solution after extraction from cereals and treatment with bicarbonate. Other metals, such as Zn and Fe, were precipitated by the bicarbonate.

III. AMINO ACIDS

Cu is complexed very readily by amino acids, peptides, and certain proteins.[7,8] Glycine and histidine are particularly important in complexing this metal. Perrin and Agarwal[9] computed the distribution of Cu complexes in human serum as follows:

Complex	% of Cu
{Cu.Cystine.His}	40
{Cu.H.Cystine.His}	31
{Cu(His)$_2$}	11
{Cu.His.SSA}	4.9
{Cu.Gln.His}	2.4
{Cu.His.Thr}	1.8
{Cu.Gly.His}	1.5
{Cu.His.Ser}	1.2

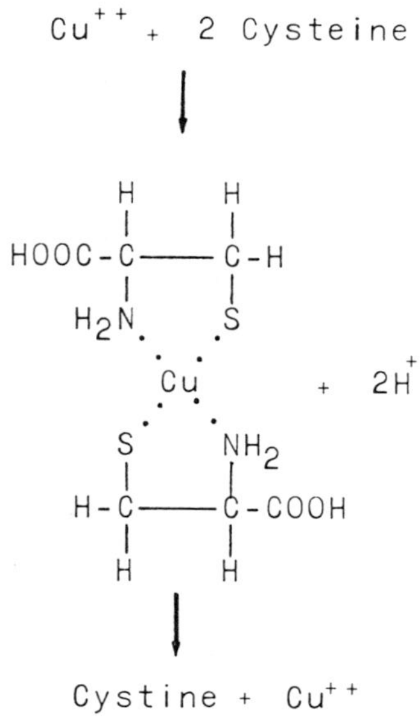

FIGURE 1. The chelation of Cu^{2+} by cysteine and subsequent oxidation to cystine. (From Robbins, K. R. and Baker, D. H., *Poult. Sci.*, 59, 1246, 1980. With permission.)

Kirchgessner and Grassman[10] reported that the in vitro dialysis of Cu sulfate was inhibited by the presence of food materials. This was explained by a decrease in the rate of diffusion by the formation of complexes, and the inhibition was increased with the extent and stability of the complexes. Fasted animals were also able to absorb Cu sulfate more rapidly than from Cu complexes. When food was given, however, macromolecular complexes were formed with the soluble Cu, which reduced the rate of absorption. When Cu was added to a rat diet complexed with an amino acid, peptide, or polypeptide complex, it was absorbed more rapidly and its level in the liver was higher than with Cu sulfate. They concluded that the stability and size of the Cu complexes and the specific effects of the ligands can play an important role in Cu absorption.

When Cu is added to a chick diet at 500 or 750 mg/kg, growth is severely depressed and hepatic and blood plasma Cu levels are greatly elevated. Jensen and Maurice[11] found that the addition of 0.4% DL-methionine or an equivalent level of cystine counteracted the growth depression and reduced the Cu levels in the blood plasma and liver. Sodium sulfate partially counteracted the Cu toxicity. Robbins and Baker[12] noted that the toxicity of Cu for chicks was greater with a purified diet containing amino acids than with one of conventional feedstuffs. They suggested that the crystalline amino acids of the purified diet might cause more rapid Cu absorption and thus greater toxicity.

Robbins and Baker[13] further investigated the effect of sulfur amino acids on Cu toxicity (Figure 1). They found that the sulfur amino acid requirement without added Cu was 0.59% but when the diet contained 500 mg of Cu per kilogram it was increased to 0.77%. Cysteine was about twice as active as cystine and other sulfur amino acid sources tested in counteracting the toxicity of Cu. The chelation of Cu by cysteine and its excretion caused a reduction of the toxic levels of Cu, but it also reduced levels of cysteine for protein synthesis. Penicillamine, which is a cysteine analog, also chelates Cu readily. When it was added to a diet marginal in sulfur amino acids, growth was depressed, but was improved again with the addition of Cu. On the other hand, Cu added to a diet marginal in cysteine resulted in a

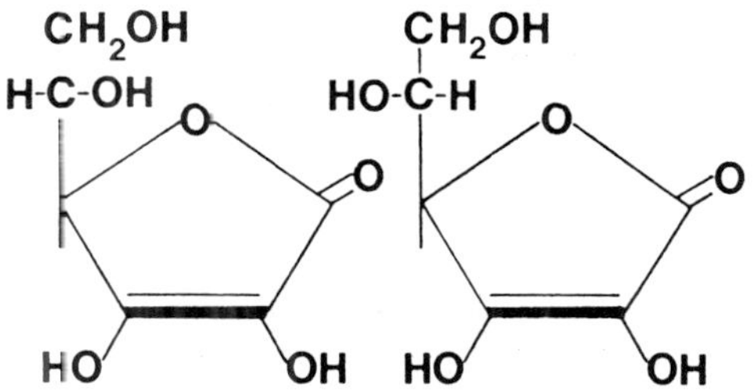

FIGURE 2. Structure of L-ascorbic acid and its analog D-isoascorbic acid. The two differ in the stereo-configuration around C-5; L-ascorbic acid is shown on the left. (From Disilvestro, R. A. and Harris, E. D., *J. Nutr.*, 111, 1964, 1981. With permission.)

growth depression due to a nutrient deficiency. This suggests that excess Cu reduces the availability of sulfur-containing free SH-groups, resulting in a growth depression as cysteine availability is reduced. Growth improvement occurs when penicillamine toxicity is decreased by Cu supplementation.

IV. ASCORBIC ACID

Ascorbic acid is antagonistic to Cu in a number of different species. Carlton and Henderson[14] reported that ascorbic acid fed to chicks at a level of 0.5% in a semipurified diet containing 8 ppm Cu caused reduced growth, lowered the hemoglobin and hematocrit values, and increased mortality. The ascorbic acid seemed to intensify the effect of a Cu deficiency, which was not alleviated by the addition of reserpine or estrogen. Hill and Starcher[15] obtained an antagonistic effect with 0.1% of dietary ascorbic acid or isoascorbic acid. The addition of 1% ascorbic acid to an adequate diet for rabbits caused a Cu deficiency and lesions were similar to those in rabbits fed a diet containing 2 ppm.[16,17] Guinea pigs fed 10 times the amount of ascorbic acid fed to the controls showed 39 and 52% lower levels of Cu in the blood and liver, respectively.[18] In a study by Milne and Omaye,[19] adult Cynomolgus monkeys fed high levels of ascorbic acid (25 mg/kg/day) with a diet marginal in Cu for a prolonged period showed small reductions in serum Cu and in serum ceruloplasmin.

The mechanism by which ascorbic acid interferes with Cu utilization has been studied by Van Campen and Gross[20] in the rat. Ascorbic acid depressed the absorption of ^{64} Cu from a ligated intestinal segment. Rats fed 1% ascorbic acid retained less of the ^{64} Cu from an oral dose, but when it was given intraperitoneally there was no difference in its retention. This indicates that ascorbic acid interferes with the absorption of Cu but has no effect on its excretion. Disilvestro and Harris[21] studied the postabsorptive effects of L-ascorbic acid on Cu metabolism in chicks (Figure 2). When administered intraperitoneally with (or 75 min before) $CuSO_4$, L-ascorbic acid significantly impaired the effectiveness of Cu to restore lysyl oxidase activity in deficient chicks. When given 75 min after the $CuSO_4$, however, there was a substantial increase in the Cu-induced enzyme activation. There was no postabsorptive stimulation with D-isoascorbic acid. This indicates antagonistic effect of ascorbic acid on Cu metabolism, apart from its influence on absorption.

The effects of ascorbic acid on Cu utilization in humans has also been studied. Finley and Cerlewski[22] studied the influence of ascorbic acid supplementation at 1.5 g/day on the Cu status of young men for a 64-day period. Serum ceruloplasmin activity was significantly reduced at every datum point during the supplementation period. There was also a trend for reduced serum Cu, but it was not significant. Both of the levels rose when the subjects returned to a control diet. This study suggests that humans respond in a similar way to experimental animals, but the reduced ceruloplasmin and serum Cu levels were still within the normal range.

V. SUGARS

There are few studies on the effect of sugars on the availability of Cu, although one might question whether the differences in the requirement for Cu which have been found with different diets might be a reflection of the carbohydrate in the experimental diet. For example, one wonders whether the high level of lactose in dried skim milk, which has been used at a level of 60% in some diets for chicks, might have an effect upon the availability of trace elements, including Cu. Hill and Matrone[23] found the chick requirement for Cu to be about 4 ppm, while a slightly higher value was reported by Davis et al.[24] with a diet based upon EDTA-washed isolated protein. A more definitive study was conducted by Reiser et al.[25] with rats. They found that fructose and sucrose increased the severity of a Cu deficiency more than starch. It was not clear whether this is a specific effect of fructose, either by itself or from sucrose, or is an effect of the size of the carbohydrate molecule. If this effect of fructose-containing carbohydrates is also true in humans, it poses a question of Cu availability in situations where the Cu level might be marginal and there is extensive use of a high-fructose corn sweetner.

VI. PROTEINS

In human milk, approximately 28% of the Cu is bound to the casein and 39% to serum albumin according to the report of Lonnerdal et al.[26] The remainder of the Cu was in a low molecular weight form or associated with the fat. Since there are differences in the availability of Zn, and possibly Cu, from human milk and cow's milk, it is thought that the specific ligands in the milk are responsible for the availability differences. It is not known whether there is redistribution of the minerals among the various binding agents during passage through the GI tract. Rivera et al.[27] consider milk to be an excellent vehicle for Fe or Cu supplementation.

VII. SYNTHETIC CHELATING AGENTS AND ANTIBIOTICS

As noted previously in relation to phytic acid, Davis et al.[28] observed a slight improvement in growth and hemoglobin formation in chickens with the addition of EDTA to a diet deficient in Cu and containing isolated soybean protein. At the other extreme, the toxicity of Cu for turkey poults was not alleviated by the addition of EDTA to the diet.[29,30]

With the thought that changes in endocrine balance might influence the serum and plasma levels of metal carrier proteins and thus influence their absorption, King et al.[31] studied the effect of oral contraceptives on the absorption of Fe, Cu, and Zn in human subjects. Blood ceruplasmin was approximately twice as high in the group of women using the oral contraceptive, but there was no statistically significant difference in the overall absorption of Cu.

Penicillamine, previously mentioned in relation to cysteine chelation of Cu, has been very useful in the treatment of Wilson's disease. This is an inherited abnormality[9,32] in which there is a low level of ceruloplasmin but excessive levels of exchangeable Cu. Penicillamine has been used very successfully in mobilizing and increasing the excretion of Cu without increasing the loss of Zn.

REFERENCES

1. **Underwood, E. J.,** *Trace Elements in Human and Animal Nutrition,* Academic Press, New York, 1977.
2. **Starcher, B. C.,** Studies on the mechanism of copper absorption in the chick, *J. Nutr.,* 97, 321, 1969.
3. **Davis, P. N., Norris, L. C., and Kratzer, F. H.,** Interferences of soy protein with the utilization of trace minerals, *J. Nutr.,* 77, 217, 1962.
4. **Lo, G. S., Settle, S. L., and Steinke, F. H.,** Bioavailability of copper in isolated soybean protein using the rat as an experimental model, *J. Nutr.,* 114, 332, 1984.
5. **Davis, P. N., Norris, L. C., and Kratzer, F. H.,** Iron deficiency studies in chicks using treated isolated soybean protein diets, *J. Nutr.,* 78, 445, 1962.
6. **Lyon, D. B.,** Studies on the solubility of Ca, Mg, Zn, and Cu in cereal products, *Am. J. Clin. Nutr.,* 39, 190, 1984.
7. **Martin, R. P., Petit-Ramel, M. M., and Scharff, J. P.,** Mixed-ligand metal ion complexes of amino acids and peptides, in *Metal Ions in Biological Systems,* Vol. 2, Sigel, H., Ed., Marcel Dekker, New York, 1973.
8. **Sigel, H.,** Structural aspects of mixed-ligand complex formation in solution, in *Metal Ions in Biological Systems,* Vol. 2, Sigel, H., Ed., Marcel Dekker, New York, 1973.
9. **Perrin, D. D. and Agarwal, R. P.,** Multimetal-multiligand equilibria: a model for biological systems, in *Metal Ions in Biological Systems,* Vol. 2., Sigel, H., Ed., Marcel Dekker, New York, 1973.
10. **Kirchgessner, M. and Grassmann, E.,** The dynamics of copper absorption, in *Trace Element Metabolism in Animals,* Proc. WAAP/IBP Int. Symp., Aberdeen, Scotland, 1969, 277.
11. **Jensen, L. S. and Maurice, D. V.,** Influence of sulfur amino acids on copper toxicity in chicks, *J. Nutr.,* 109, 91, 1979.
12. **Robbins, K. R. and Baker, D. H.,** Effect of high-level copper feeding on the sulfur amino acid need of chicks fed corn-soybean meal and purified crystalline amino acid diets, *Poult. Sci.,* 59, 1099, 1980.
13. **Robbins, K. R. and Baker, D. H.,** Effect of sulfur amino acid level and source on the performance of chicks fed high levels of copper, *Poult. Sci.,* 59, 1246, 1980.
14. **Carlton, W. W. and Henderson, W.,** Studies in chickens fed a copper deficient diet supplemented with ascorbic acid, reserpine and diethylstilbestrol, *J. Nutr.,* 85, 67, 1965.
15. **Hill, C. H. and Starcher, B.,** Effect of reducing agents on copper deficiency in the chick, *J. Nutr.,* 85, 271, 1965.
16. **Hunt, C. E. and Carlton, W. W.,** Cardiovascular lesions associated with experimental copper deficiency in the rabbit, *J. Nutr.,* 87, 385, 1965.
17. **Hunt, C. E., Carlton, W. W., and Newberne, P. M.,** Interrelationships between copper deficiency and dietary ascorbic acid in the rabbit, *Br. J. Nutr.,* 24, 61, 1970.
18. **Smith, C. H. and Bidlack, W. R.,** Interrelationships of dietary ascorbic acid and iron on the tissue distribution of ascorbic acid, iron and copper in female guinea pigs, *J. Nutr.,* 110, 1398, 1980.
19. **Milne, D. B. and Omaye, S. T.,** Effects of vitamin C on copper and iron metabolism in the guinea pig, *Int. J. Vitam. Nutr. Res.,* 50, 301, 1980.
20. **Van Campen, D. R. and Gross, E.,** Influence of ascorbic acid on the absorption of copper by rats, *J. Nutr.,* 95, 617, 1968.
21. **Disilvestro, R. A. and Harris, E. D.,** A postabsorptive effect of L-ascorbic acid on copper metabolism in chicks, *J. Nutr.,* 111, 1964, 1981.
22. **Finley, E. G. and Cerklewski, F. L.,** Influence of ascorbic acid supplementation on copper status in young adult men, *Am. J. Clin. Nutr.,* 37, 553, 1983.
23. **Hill, C. H. and Matrone, G.,** Studies on copper and iron deficiencies in growing chickens, *J. Nutr.,* 73, 425, 1961.
24. **Davis, P. N., Norris, L. C., and Kratzer, F. H.,** Iron deficiency studies in chicks using treated isolated soybean protein diets, *J. Nutr.,* 78, 445, 1962.
25. **Reiser, S., Ferretti, R. J , Fields, M., and Smith, J. C.,** Role of dietary fructose in the enhancement of mortality and biochemical changes associated with copper deficiency in rats, *Am. J. Clin. Nutr.,* 38, 214, 1983.
26. **Lonnerdal, B., Hoffman, B., and Hurley, L. S.,** Zinc and copper binding proteins in human milk, *Am. J. Clin. Nutr.,* 36, 1170, 1982.
27. **Rivera, R., Ruiz, R., Hegenauer, J., Saltman, P., and Green, R.,** Bioavailability of iron- and copper-supplemented milk for Mexican school children, *Am. J. Clin. Nutr.,* 36, 1162, 1982.
28. **Davis, P. N., Norris, L. C., and Kratzer, F. H.,** Interference of soy protein with the utilization of trace minerals, *J. Nutr.,* 77, 217, 1962.
29. **Vohra, P., Gottfredson, G. D., and Kratzer, F. H.,** The effects of high levels of dietary EDTA, zinc or copper on the mineral contents of some tissue of turkey poults, *Poult. Sci.,* 47, 1334, 1968.
30. **Vohra, P. and Kratzer, F. H.,** Zinc, copper and manganese toxicities in turkey poults and their alleviation by EDTA, *Poult. Sci.,* 47, 699, 1968.

31. **King, J. C., Raynolds, W. L., and Margen, S.,** Absorption of stable isotopes of iron, copper and zinc during oral contraceptive use, *Am. J. Clin. Nutr.*, 31, 1198, 1978.
32. **Scheinberg, I. H.,** Wilson's Disease and the physiological chemistry of copper, in *Inorganic Chemistry in Biology and Medicine,* Martell, A. E., Ed., A.C.S. Symp. Series 140, Washington, D.C., 1980.
33. **Perrin, D. D. and Agarwal, R. P.,** Multimetal-multiligand equilibria: a model for biological systems, in *Metal Ions in Biological Systems,* Vol. 2, Sigel, H., Ed., Marcel Dekker, New York, 1973.

Chapter 9

OTHER MINERALS

I. Ca

The absorption of Ca and factors which influence it have been reviewed by Allen.[1] Both active transport and diffusion are important in the absorption process. It has been estimated that with a low-Ca diet as much as 95% of the Ca is absorbed by active transport; this figure is greatly reduced as the Ca in the diet is increased. A number of plant components influence the availability of calcium (Table 1).

During World War II it was observed in England that Ca was poorly used with the consumption of bread with 92% extraction of wheat flour in comparison to bread with 69% extraction in which more of the outer fraction of the wheat berry was removed.[2] Sodium phytate added to the 69% extraction flour depressed the absorption of Ca, indicating that phytate in the bran might be responsible for the poor Ca utilization.

As with other minerals, the role of phytic acid or the fiber in the bran fraction to influence Ca availability is still not clear. Reinhold et al.[3] found that the consumption of tanok, a phytate-rich unleavened whole-meal flat bread, for 32 days caused strongly negative Ca balances in young male subjects. The effect of phytate was less definite when leavened flat bread was supplemented with 2.5 g of phytic acid daily. In this case, Ca balance was negative in one subject, marginal in another, and unchanged in a third. Slavin and Marlett[4] fed a low-fiber diet supplemented with 16 g of refined cellulose daily and found that the fecal excretion of Ca was increased and there was a negative Ca balance. James et al.[5] found that dietary fiber from plants low in phytate-bound Ca in proportion to its uronic acid content (Figure 1). A typical vegetarian diet may contain sufficient uronic acids to bind about 360 mg of dietary Ca per day,[6] but over 80% of the acids may be fermented in the intestine. The noncellulose fraction of the fiber reduces the amount of Ca available for absorption by the small intestine, but the Ca might be released in the colon by microbial digestion of the uronic acids and be available for colonic absorption. It may be possible then, to maintain Ca balance on a high-fiber diet after an adaptive period.

Carrageenan and agar-agar reduced the absorption of Ca in short-term studies with rats,[7] but in a long-term study, 10% of agar-agar did not cause any difference in the Ca content of the rat carcass. It may be that a part of the adaption process may involve a reduction in the urinary excretion of Ca as was shown in adult male chickens fed a diet very low in Ca.[8] Pectin, which also contains a high level of uronic acids, had no effect on Ca balance when consumed at 36 g/day by young men for a 6-week period.[9] Again, pectin is metabolized in the lower digestive tract, which would free the Ca for absorption.

The level of oxalic acid in food has been a nutritional concern since calcium oxalate is extremely insoluble. Zarembski and Hodgkinson[10] found the mean daily intake of oxalic acid to range from 70 to 150 mg/day in English diets with tea being the largest single source. Spinach is a notoriously rich source of oxalic acid. It is questionable whether oxalic acid from one foodstuff will affect the availability of Ca from another source. Studies with chickens[11] indicated that one should consider the total oxalate as well as the Ca content of the diet from a nutritional standpoint. Dehydrated beet leaves (4.26% oxalic acid) at 10% of the diet depressed the growth of chickens when the ration contained 0.9% Ca, but increasing the Ca overcame the depressing effect. Similarly, oxalic acid at 1% of the diet depressed growth, bone ash, and breaking strength at 0.9% Ca, but the effects were corrected by increasing the Ca.

Lactose is known to enhance Ca utilization. It increases the diffusional component of the

Table 1
EFFECT OF PLANT COMPONENTS ON Ca ABSORPTION

Phytate	Decrease
Cellulose	Decrease
Uronic acid	Decrease
Pectin	Null
Sodium alginate	Decrease
Oxalate	Decrease
Ascorbic acid	Null

From Allen, L. H., *Am. J. Clin. Nutr.*, 35, 783, 1982. With permission.

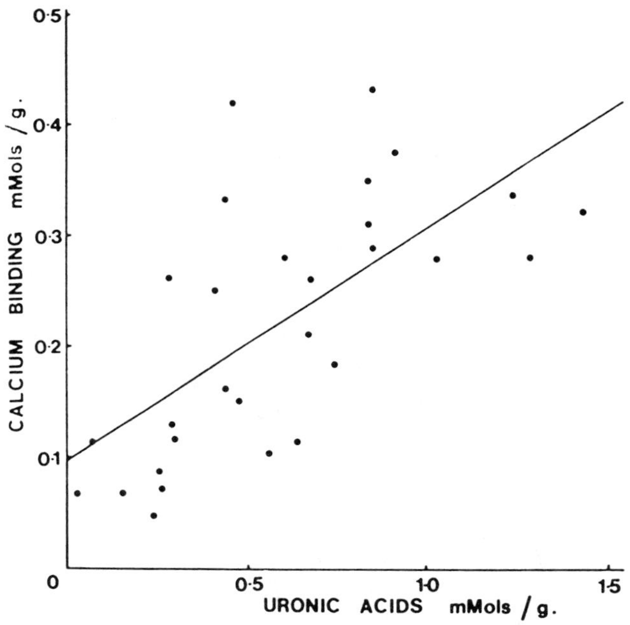

FIGURE 1. Relationship of Ca binding to uronic acid content of material extracted from different plant foods. Ca binding was assessed at pH 7.4. Regression equation for binding in relation to uronic-acid content was y = 0.209x + 0.097 with a correlation coefficient of 0.66. (From James, W. P. et al., *Lancet*, 1, 638, 1980. With permission.)

transport system rather than the active transport component, which is dependent upon vitamin D. Lengemann et al.[11] found improved Ca transport with lactose within 30 min of introduction into the ileum and that there was an appreciably greater response in the ileum than the duodenum or jejunum. In further studies, Armbrecht and Wasserman[13] used everted gut sacs from rat ileum which were pre-incubated with or without lactose for 45 min (Figure 2). When Ca^{2+} uptake during a 45-min period was measured in parallel experiments, no evidence for the cotransport of lactose and Ca^{2+} into the tissue was found. These and other data indicate that lactose is not absorbed directly with the Ca, but may affect the absorptive cells of the intestine to increase their permeability to Ca.

Ca absorption in humans is also enhanced by lactose. Kobayashi et al.[14] studied Ca and

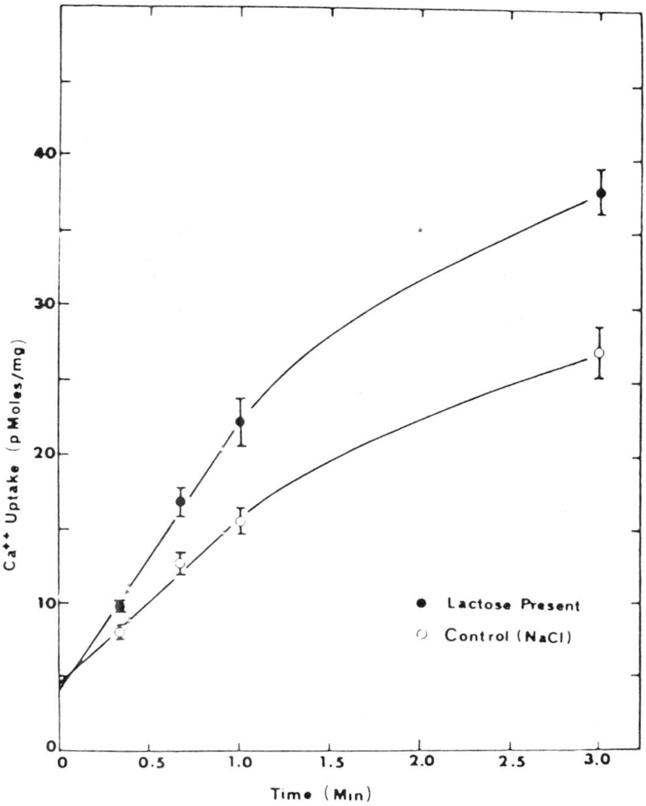

FIGURE 2. Uptake of 0.1 m*M* Ca^{2+} with time by rat ileum. Lactose (160 m*M*) was present or absent during both 45-min pre-incubation and actual measurement of Ca^{2+} uptake for indicated time. Each point is the mean ± SEM of 12 gut sacs and the lactose values are significantly greater than the control at all time points ($p < 0.005$). (From Armbrecht, H. J. and Wasserman, R. H., *J. Nutr.*, 106, 1265, 1976. With permission.)

Mg absorption in infants between the ages of 2 and 8 months (Figure 3). Ca absorption was about 60% in infants supplied proprietary milk (Group A), was reduced to 36% when lactose-free milk was supplied (Group B), and was increased to 72% when a lactase preparation was added to the proprietary milk (Group C). These results suggest that the effect of lactose in enhancing the absorption of Ca may be due to the effect of the constituent sugars, glucose, and galactose, since the addition of lactase was even more effective than lactose alone. The absorption of Mg paralleled that of Ca. In rat studies,[15] lactose, sorbitol, glucose, sucrose, fructose, galactose, and xylose did not significantly influence Ca absorption from the duodenum; however, each of the sugars greatly enhanced the absorption of Ca from the ileum. These data suggest that dietary lactose may enhance Ca absorption because it is slowly absorbed in the upper intestinal tract and its constituent sugars are available to aid in Ca absorption in the lower intestinal region.

There are four main explanations which have been proposed for the enhancing effect of lactose on Ca absorption. The first theory envisions the chelation of Ca by lactose to increase its absorption in complex form. The second proposal is that the fermentation of lactose in the intestinal tract reduces the pH of the tract, which promotes Ca solubility and absorption. A third theory proposes that lactose has a specific metabolic effect on the mucosal cells to increase the permeability to Ca. The last theory proposes that the constituent sugars from

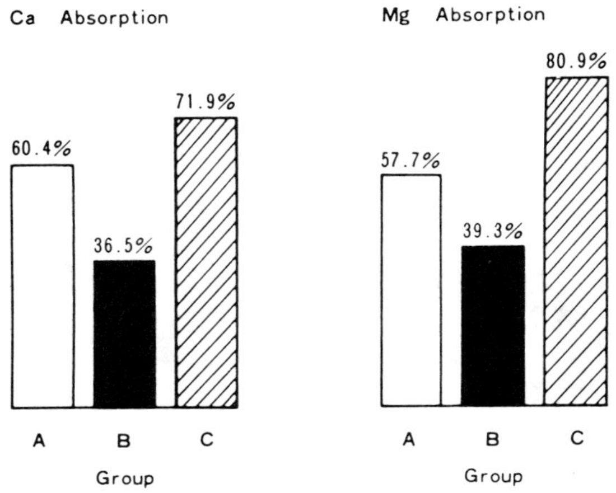

FIGURE 3. Mean value of percent absorption of Ca and Mg in infants in Groups A (milk), B (lactose-free milk), and C (milk plus lactase). (From Kobayashi, A. et al., *Am. J. Clin. Nutr.*, 28, 681, 1975. With permission.)

lactose hydrolysis may in some way stimulate Ca absorption. The specific effects of glucose and galactose may depend upon their ability to chelate Ca, either for transport to the brush border or through the mucosa. The last two mechanisms seem the most plausible at present.

II. Mn

In the classical work demonstrating the need for Mn in rations for chicks,[16] this element was noted as a contaminant in a Ca-phosphate supplement which was found to prevent rather than aggravate an enlarged hock condition. Although somewhat analogous to the effect of Ca in reducing the availability of Zn, it is probable that the effect on Mn availability is due to its coprecipitation with Ca and P in the intestinal tract. Chelation with bile constituents in the intestinal tract probably aids absorption.

Davis et al.[17] studied the effect of EDTA on growth and the severity of leg abnormalities in chicks fed diets containing isolated soybean protein and supplemented with various levels of Mn. Although the differences were not as striking as when Zn was studied, there was definite evidence that 0.07% EDTA improved growth and reduced the severity of perosis when added to a diet marginal in Mn.

In their studies of ^{54}Mn absorption in chicks with the use of the whole body counting technique, Suso and Edwards[18] found a slight improvement, although statistically insignificant, in Mn absorption with the addition of EDTA to the diet up to 0.18%, but it was much less than for Zn. There was also a slight increase in absorption as the stability constants of the chelating agents used increased up to 15.11 (DTPA), very similar, but to a much lesser extent than was noted for Zn. There are differences between human milk, cow's milk, and infant formula in the ligands which bind Mn.[19,20] Some of the ligands are proteins with molecular weights in the range of 80,000 to 407,000, but some low molecular weight Mn-binding ligands are also present in cow's milk and infant formula. The intestinal absorption of Mn in the rat was studied by an in vivo perfusion system.[21] (See Figure 4.) Mn in the presence of either L-histidine or citrate was absorbed more rapidly from the ileum or jejunum than when no ligands were included in the Mn-containing perfusion solution. The kinetics of the absorption process were calculated and were compatible with a high-affinity, low-capacity, active transport mechanism for Mn absorption in the rat intestine. This suggested

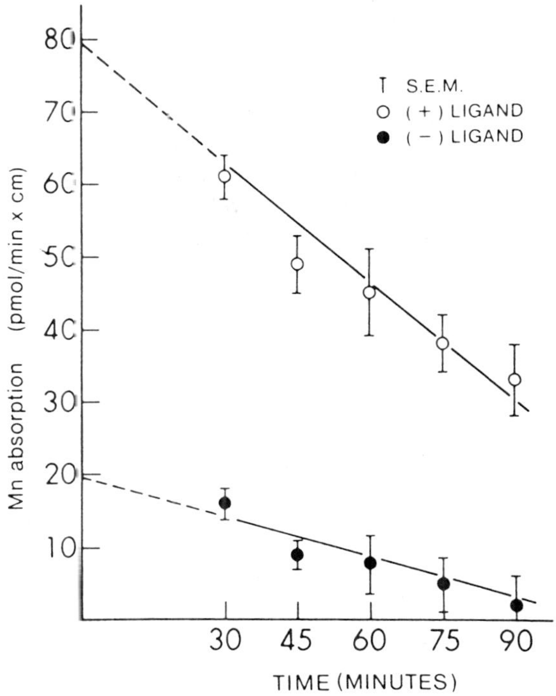

FIGURE 4. Jejunal absorption rates of Mn in the rat as a function of time. All solutions contained 0.0125 m*M* Mn with or without 0.025 m*M* of citrate as ligand. The dashed extrapolation of the regression lines intersect the ordinates at the estimated initial absorption rate. (From Garcia-Aranda, J. A. et al., *J. Nutr.*, 113, 2601, 1983. With permission.)

a limited role for small molecular weight ligands associated with both diffusion and active transport.

III. Mo

Reid et al.[22] found that the addition of distiller's dried solubles to a basal diet containing glucose and isolated soybean protein gave a growth response in chickens and poults. The ash of this material also gave a growth response which could be duplicated by the addition of molybdic acid. An interesting aspect of the response was the fact that the basal diets contained 0.972 and 1.58 ppm of Mo, respectively, for the chicks and poults, and the level of molybdic acid added was only 0.0126 ppm. This suggested that much of the Mo in the basal diet may have been nutritionally unavailable. Richert and Westerfeld[23] indicated that Mo is largely unavailable for the rat in isolated soybean protein.

Leach et al.[24] observed a growth response to Mo in chicks fed a diet containing repurified isolated soybean protein, but no response was noted with a casein diet which had a much lower Mo content than the isolated soybean protein diet. In addition, less tungsten, a known Mo antagonist, was required to antagonize an equivalent amount of Mo when the repurified isolated soybean protein diet was fed than found by other workers who used a vitamin-free casein diet.

Payne[25] described a Mo-responsive condition in growing chickens in Australia, although the Mo level in the feed was usually in excess of the required level. He pointed out the effects that other ions, such as tungstate, fluoride, sulfate, and Cu, can have on the Mo requirement as well as Cd, Mn, Zn, and Fe which can interact indirectly.

These results suggest strongly that Mo in isolated soybean protein is relatively unavailable, but the reason for its poor availability is not known. It seems unlikely that the effect can all be due to the interaction of the inorganic constituents of the soybean protein. It also seems unlikely that phytic acid can be responsible for the unavailability since Mo is present in the anion form. On the other hand, the Mo level in isolated soybean protein is reduced by repeated extraction with an EDTA solution.[26]

IV. Co

Turk and Kratzer[27] found that Co is more toxic to chickens when added to a purified diet than to a practical diet of conventional feedstuffs. Toxicity symptoms were emaciation, debility, inanition, sensitivity to being touched, poor growth, and mortality. Alleviation of the toxicity of Co was obtained by the addition of EDTA, cystine, or cysteine, but not methionine. This indicates that it is a chelation phenomenon which binds the Co to reduce its toxicity for the chick. It is probable that the partial protection from the diet of conventional feedstuffs was due to a component of the diet which has Co-binding properties.

Co was included in the studies of Suso and Edwards[18] in which the absorption of several minerals was determined by the whole body counting technique after administration of the radioactive isotope to chickens. Approximately 13% of the Co was retained when the Co was administered alone and it decreased progressively to about 7 or 8% with the addition of EDTA up to 0.24% of the diet. This agrees very well with the feeding results of Turk and Kratzer[27] in showing that EDTA makes Co less available and thus, less toxic to the chicken.

V. V

Romoser et al.[28] studied the effect of adding V to a corn-soybean meal diet and found that the addition of up to 20 mg of V per kilogram of diet was tolerated by chickens. Increasing the level to 30 mg/kg resulted in significant growth depression, but no mortality was noted until 200 mg or more of V was added per kilogram of diet. Hathcock et al.[29] studied V toxicity in chickens using a diet containing 60% dried skim milk and 32.5% glucose. The addition of 25 mg of V per kilogram of diet caused a significant decrease in growth rate and nearly 100% mortality at 2 weeks of age. It appeared that V was much more toxic when added to the skim milk-glucose diet than to the corn-soybean meal diet. The addition of EDTA at two to six times the molecular concentration of V practically eliminated the toxic symptoms.

Berg[30] also showed that the composition of the diet had a marked effect on V toxicity in the chick. He used a diet with 20 mg of V added per kilogram of diet and observed practically no growth depression with a corn-herring meal diet, while a depression of approximately 50% occurred with a sucrose-herring meal diet. A corn-soybean diet caused a growth depression of approximately 25 to 30%. The effects of the various feed ingredients were not due to differences in V content. Corn fractions, ash, oil, starch, zein, and gluten did not reduce V toxicity.

In further studies, Berg and Lawrence[31] found that the replacement of 5% sucrose in a sucrose-fishmeal diet for chicks with degossypolized cottonseed meal prevented the toxicity of 20 mg of V per kilogram of diet. The addition of 5% dehydrated grass to the same diet also greatly reduced the toxicity of V, but the addition of soybean meal, corn gluten meal, meat meal, fish meal, casein, isolated soybean protein, zein, and wheat gluten had no effect. The addition of 0.25 to 0.5% ascorbic acid to the sucrose-fishmeal diet prevented the toxicity

Table 2
THE EFFECT OF LACTOSE ON VANADIUM TOXICITY

Treatment	Average 4 weeks body weight gain (g) ± SEM	Mortality (%)	% Growth depression	Significance of difference from basal
Basal[a]	490.3 ± 25.0	0	0	
Basal + 100 ppm V	289.1 ± 42.2	0	41.0	$p < 0.01$
Basal + 10% lactose	446.3 ± 22.2	0	9.0	NS
Basal + 100 ppm V + 10% lactose	116.7 ± 25.5	80	76.2	$p < 0.01$

[a] The diet composition is shown in Table 1.

From Hafez, Y. S. M. and Kratzer, F. H., *Poultry Sci.*, 55, 918, 1976. With permission.

of 20 mg of V added as NH_4VO_3 per kilogram of diet. The toxicity of V from VOSO and $VOCl_2$ was the same as from NH_4VO_3, which suggested that the effect of ascorbic acid was not related to its reducing effect upon the V salts.

Hafez and Kratzer[32] found V to be more toxic to chickens when added to a semipurified diet than to a practical diet. EDTA at six times the equivalent molar level of V greatly reduced its toxicity. Lactose at 10% of the diet (Table 2) greatly increased the toxicity of V as measured both by the depression of growth and the increase in mortality. In the same study, the addition of Cr to a diet containing V greatly reduced the toxicity of V.

It is clear that the toxicity of V is greatly influenced by the diet in which it is fed. A purified diet or the addition of skim milk, sucrose, or lactose to the diet increases the toxicity of V. It is probable that the effect of skim milk is due to its lactose component. Ascorbic acid and EDTA, on the other hand, reduce the toxicity of V. While these are distinct differences in the availability of the element, it is difficult to explain these positive and negative effects of chelation. The effect of Cr in reducing the toxicity of V is predictable from the Hill and Matrone[33] proposal that "those elements whose physical and chemical properties are similar will act antagonistically to each other biologically".

VI. Au

Although not normally a dietary constituent, Au may be considered in relation to chelation since it has been used in the treatment of rheumatoid arthritis. Rubin et al.[34] studied the effect on Au metabolism in rats of five different chelating agents: DL-penicillamine, thiomalic acid, mercaptoethyliminodiacetic acid (MAIDA), 1,8-diamino-3,6-dimercaptooctane-*N,N'*-tetraacetic acid (BATA), and 2,3-dimercaptopropanol (BAL). They found that the tissue distribution and excretion of injected Au was dependent upon the Au carrier. DL-Penicillamine was the most effective of a group of –SH containing Au-binding compounds studied (Figures 5 and 6), although all of the chelating agents except MAIDA increased the urinary excretion of Au after administering them to rats which had been previously injected with Au thioglucose. Many of the chelating agents also caused a decrease in the Au content of the spleen and lung and an increased Au deposition in the liver and kidney during the mobilization of the predeposited tissue Au.

It is probable that the effectiveness of the chelating agents in reducing the toxicity of Au is due both to the increased excretion of Au and the redistribution of Au to tissues, where its presence is less critical. The time interval between the deposition of the Au and the administration of the chelating agent was also important in determining the effectiveness of the chelating agent to mobilize the Au.

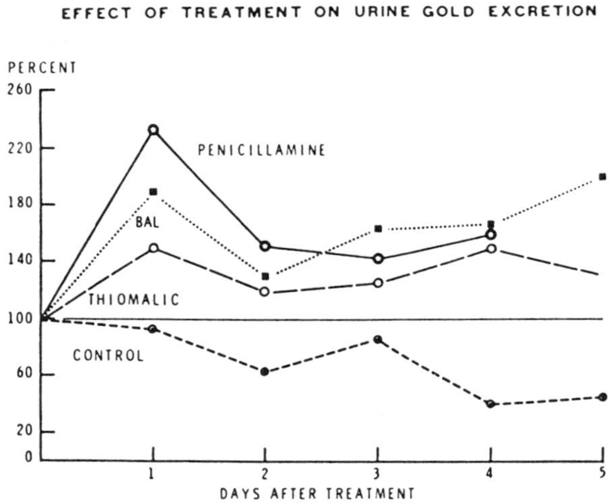

FIGURE 5. Effect of chelate treatment on urinary Au administered as Au thioglucose. (From Rubin, M. et al., *Proc. Soc. Exp. Biol. Med.*, 124, 290, 1967. With permission.)

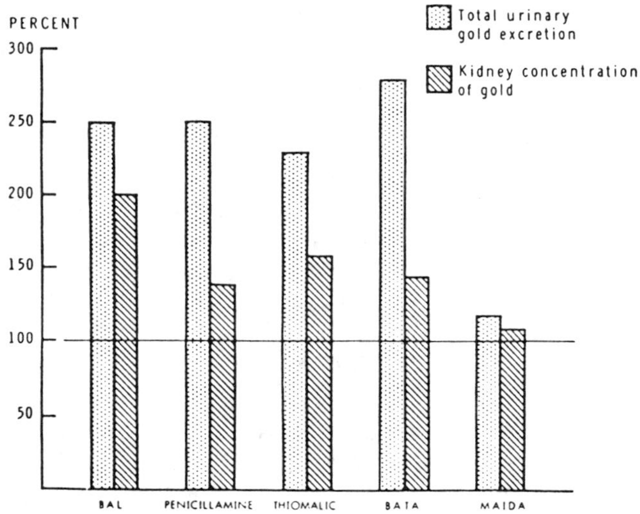

FIGURE 6. Relation of urinary and kidney Au concentration after chelate treatment compared to untreated controls (100%) after Au thioglucose administration. (From Rubin, M. et al., *Proc. Soc. Exp. Biol. Med.*, 124, 290, 1967. With permission.)

REFERENCES

1. **Allen, L. H.,** Calcium bioavailability and absorption: a review, *Am. J. Clin. Nutr.*, 35, 783, 1982.
2. **McCance, R. A. and Widdowson, E. M.,** Mineral metabolism of healthy adults on white and brown bread dietaries, *J. Physiol.*, 101, 44, 1942.
3. **Rheinhold, J. G., Nasr, K., Lahimgarzadeh, R., and Hedayati, H.,** Effects of purified phytate and phytate-rich bread upon metabolism of zinc, calcium, phosphorus and nitrogen in man, *Lancet,* 1, 283, 1973.

4. **Slavin, J. L. and Marlett, J. A.,** Influence of refined cellulose on human small bowel function and calcium and magnesium balance, *Am. J. Clin. Nutr.*, 33, 1932, 1980.
5. **James, W. P. T., Branch, W. J., and Southgate, D. A. T.,** Calcium binding by dietary fiber, *Lancet*, 1, 638, 1978.
6. **Branch, W. J., Southgate, D. A. T., and James, W. P. T.,** Binding of calcium by dietary fiber: its relationship to unsubstituted uronic acids, *Proc. Nutr. Soc.*, 34, 120A, 1975.
7. **Harmuth-Hoene, A. S. and Schelenz, R.,** Effect of dietary fiber on mineral absorption in growing rats, *J. Nutr.*, 110, 1774, 1980.
8. **Norris, L. C., Kratzer, F. H., Lin, H. J., Hellewell, A. B., and Beljan, J. R.,** Effect of quantity of dietary calcium on maintenance of bone integrity in mature White Leghorn male chickens, *J. Nutr.*, 102, 1085, 1972.
9. **Cummings, J. H., Southgate, D. A. T., Branch, W. J., Wiggins, H. S., Houston, H., and Jenkins, D. J. A.,** The digestion of pectin in the human gut and its effect on calcium absorption and large bowel function, *Br. J. Nutr.*, 41, 477, 1979.
10. **Zarembski, P. M. and Hodgkinson, A.,** The oxalic acid content of English diets, *Br. J. Nutr.*, 16, 627, 1962.
11. **Lengemann, F. W., Wasserman, R. H., and Comar, C. L.,** Studies on the enhancement of radioCa and radioSr absorption by lactose in the rat, *J. Nutr.*, 68, 443, 1959.
12. **Robertson, E. I., Brin, M., and Norris, L. C.,** The use of dehydrated beet leaves in chick rations, *Poult. Sci.*, 26, 582, 1947.
13. **Armbrecht, H. J. and Wasserman, R. H.,** Enhancement of Ca^{++} uptake by lactose in the rat small intestine, *J. Nutr.*, 106, 1265, 1976
14. **Kobayashi, A., Kawai, S., Ohbe, Y., and Nagashima, Y.,** Effects of dietary lactose and a lactase preparation on the intestinal absorption of calcium and magnesium in normal infants, *Am. J. Clin. Nutr.*, 28, 681, 1975.
15. **Vaughn, O. W. and Filer, L. J., Jr.,** The enhancing action of certain carbohydrates on the intestinal absorption of calcium in the rat, *J. Nutr.*, 71, 10, 1960.
16. **Wilgus, H. S., Jr., Norris, L. C., and Heuser, G. F.,** The role of certain inorganic elements in the cause and prevention of perosis, *Science*, 84, 252, 1936.
17, **Davis, P. N., Norris, L. C., and Kratzer, F. H.,** Interference of soy protein with the utilization of trace minerals, *J. Nutr.*, 77, 217, 1962.
18. **Suso, F. A. and Edwards, H. M., Jr.,** Influence of various chelating agents on absorption of ^{60}Co, ^{59}Mn and ^{65}Zn by chickens, *Poult. Sci.*, 47, 1417, 1968.
19. **Chan, W.-Y., Bates, J. M., Jr., and Reussert, O. M.,** Comparative studies of manganese binding in human breast milk, bovine milk and infant formula, *J. Nutr.*, 112, 642, 1982.
20. **Lonnerdal, B., Keen, C. L., and Hurley, L. S.,** Manganese binding in human milk and cow's milk — an effect on bioavailability, *Fed. Proc. Fed. Am. Soc. Exp. Biol.*, 42, 926, 1983.
21. **Garcia-Aranda, J. A., Wapnir, R. A., and Lifshitz, F.,** In vivo intestinal absorption of manganese in the rat, *J. Nutr.*, 113, 2601, 1983.
22. **Reid, B. L., Kurnick, A. A., Svacha, R. L., and Couch, J. R.,** The effect of molybdenum on chick and poult growth, *Poult. Sci.*, 93, 245, 1956.
23. **Richert, D. A. and Westerfeld, W. W.,** Isolation and identification of the xanthine oxidase factor as molybdenum, *J. Biol. Chem.*, 203, 915, 1953.
24. **Leach, R. J., Turk, D. E., Zeigler, T. R., and Norris, L. C.,** Studies on the role of molybdenum in chick nutrition, *Poult. Sci.*, 41, 300, 1962.
25. **Payne, C. G.,** Involvement of molybdenum in feather growth, *Br. Poult. Sci.*, 18, 427, 1977.
26. **Davis, P. N., Norris, L. C., and Kratzer, F. H.,** Iron deficiency studies in chicks using treated isolated soybean protein diets, *J. Nutr.*, 78, 445, 1962.
27. **Turk, J. L., Jr. and Kratzer, F. H.,** The effects of cobalt in the diet of the chick, *Poult. Sci.*, 39, 1302, 1960.
28. **Romoser, G. L., Dudley, W. A., Machlin, L. J., and Loveless, L.,** Toxicity of vanadium and chromium for the growing chick, *Poult. Sci.*, 40, 1171, 1961.
29. **Hathcock, J. N., Hill, C. H., and Matrone, G.,** Vanadium toxicity and distribution in chicks and rats, *J. Nutr.*, 82, 106, 1964.
30. **Berg, L. R.,** Effect of diet composition on vanadium toxicity for the chick, *Poult. Sci.*, 45, 1346, 1966.
31. **Berg, L. R. and Lawrence, W. W.,** Cottonseed meal, dehydrated grass and ascorbic acid as dietary factors preventing toxicity of vanadium for the chick, *Poult. Sci.*, 50, 1399, 1971.
32. **Hafez, Y. S. M. and Kratzer, F. H.,** The effect of diet on the toxicity of vanadium, *Poult. Sci.*, 55, 918, 1976.
33. **Hill, C. H. and Matrone, G.,** Chemical parameters in the study of in vivo and in vitro interactions of transition elements, *Fed. Proc. Fed. Am. Soc. Exp. Biol.*, 29, 1474, 1970.
34. **Rubin, M., Sliwinski, A., Photias, M., Feldman, M., and Zvaifler, N.,** Influence of chelation on gold metabolism in rats, *Proc. Soc. Exp. Biol. Med.*, 124, 290, 1967.

Chapter 10

REPRODUCTION

I. INTRODUCTION

The fetus requires a continuous supply of nutrients for it growth and metabolism,[1] with a rather high proportion of the nutrients used for tissue formation. The transfer across the placenta takes place by three processes: diffusion, facilitated diffusion, and active transport. In many caes it is difficult to identify the process because a greater concentration of organic nutrients in the fetal circulation than in the maternal circulation may indicate the synthesis by the fetus, instead of transport against a concentration gradient. An example of this is the presence of fructose in the fetal plasma at a much higher concentration than in the maternal plasma, due to production from glucose in the fetus.

The levels of mineral in the fetal and maternal circulations cannot be complicated by synthesis. However, the presence of a mineral in free and bound forms may make the interpretation of total concentration data difficult. For example, the higher fetal than maternal levels of Ca might seem to imply active transport; one must first demonstrate that the concentrations of nonprotein-bound Ca of the fetal plasma is higher than the maternal plasma. The effect of chelating agents on the transport of minerals across the placenta has not been studied extensively.

II. Fe

The problem of Fe accumulation by the fetus is especially important since young animals are frequently deficient in Fe.[2] Normally, plasma Fe is delivered to the placenta as maternal Fe transferrin (Figure 1). After passage through the placenta, it is made available to the fetal circulation, where it is bound by the fetal transferrin. It is then used for fetal storage and hemoglobin formation with no transfer back to the maternal circulation.

With appropriate surgical procedures, it is possible to introduce Fe in several forms into the maternal compartment and sample the perfusate from the fetal side of the placenta. In studies with guinea pigs, Bagley et al.[3] studied the effect of several chelating agents on the transplacental passage of maternally administered Fe to the fetal circulation (Figure 2). The placental barrier was impermeable to ferric citrate, ferric fructose, a number amino acid chelates, and ferrioxamine Fe chelates. Fe chelated with fructose has enhanced intestinal absorption, but was not transferred across the placenta any better than the free Fe. Any effect of fructose fed to the mother on the Fe content of the young would be due to improved intestinal absorption, rather than improved placental transfer. Direct placental passage was observed of administered Fe bound to *N,N'*-ethylene-*bis*-(2-[2-hydroxy-5-cholorphenyl] glycine (Cl-EHPG), and to a lesser extent, to EHPG and DTPA. These compounds apparently bypass the normal Fe transport mechanisms of the placental barrier. Intact radioiron Cl-EHPG was still detectable in the plasma of the fetus 90 min post-injection, but the majority of the radioiron was associated with the fetal transferrin. This indicates that the Fe, although absorbed as the chelate, is readily available nutritionally.

III. Cu

The crinkled (cr) mutant mouse has been shown to have developmental abnormalities in Cu transport and storage.[4] Dietary supplementation of the dam with Cu sulfate during pregnancy and lactation will ameliorate partially the effects of the gene. Cu nitrilotriacetate

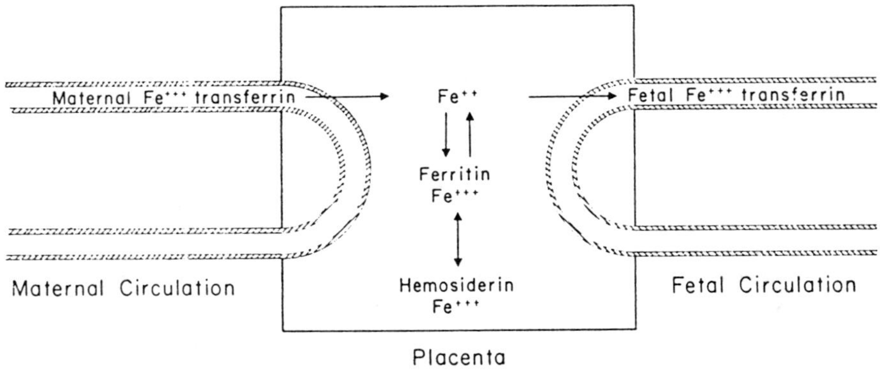

FIGURE 1. Generalized compartmental distribution of iron. (From Rubin, M., *Am. Feed. Manufact. Symp.*, 27, 1968. With permission.)

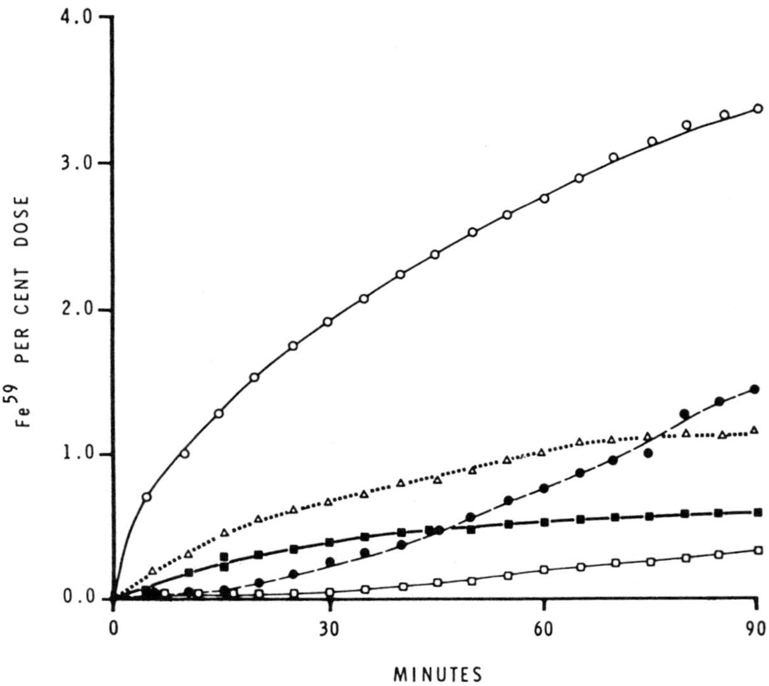

FIGURE 2. Cumulative appearance of ^{59}Fe in placental perfusate. Dose: 6 mg Fe/kg iv. Samples 3.5 mℓ each 5 min. ○, *N,N'*ethylene-*bis*- (2 [2-hydroxy-5-chlorophenyl] glycine) (Cl-EHPG);●, Plasma, ▲ diethylenetriaminepenta-acetic acid (DTPA); ■, *N,N'*ethylene-bis-(2-[O-hydroxy-phenyl] glycine) (EHGP); □, other compounds: *N,N'*ethylene-*bis*-(2-[2-hydroxy-5-sulfophenyl] glycine) (SO_3H-EHPG); 1,2-diamino-cyclohexane-*N,N'*-tetra-acetic acid (CDTA); ferrioxamine; ferrous and ferric fructose; and citrate.

(NTA) was shown to be more effective than copper sulfate in increasing postnatal survival and body Cu content of offspring when dams were supplemented during pregnancy and lactation. Feeding NTA alone had no effect on survival. The postnatal supplementation with Cu did not increase the survival of the mutants, showing that prenatal treatment was essential. There are similarities between the cr condition in the mutant mouse and Menkes' kinky hair syndrome. This is a genetic disorder of humans produced by a sex-linked recessive gene which causes the synthesis of an abnormal protein with high Cu-binding activity. Cu supplementation after birth is ineffective, but Keen et al.[4] suggested that prenatal Cu · NTA supplementation may be as effective as it has been in the mutant mouse.

D-Penicillamine (DP) has the capacity to chelate metals, particularly Cu, and increases the rate of Cu excretion in the urine. It has been used for the treatment of a number of disorders, including Wilson's disease, a genetic disorder characterized by high levels of Cu in the tissue. Several cases of malformations in infants whose mothers had taken DP throughout pregnancy suggested a deficiency of Cu and prompted Keen et al.[5] to study the teratogenicity of DP in rats. The frequency of resorptions and the frequency and severity of malformations increased with increasing levels of the drug. Maternal and fetal tissue Cu levels decreased with increasing DP levels and, in addition, fetal Zn levels were lower in the 1.6% DP group than in the controls. The results suggested that the teratogenicity of DP may be due to induction of a Cu deficiency and, at high levels, also of a Zn deficiency. In further studies,[6] it was shown that increasing the level of Cu in a diet for rats containing DP greatly reduced the frequency of resorptions and malformations. Cu levels were significantly reduced in both the maternal and fetal tissues, so one might explain the effect of DP as creating a Cu deficiency in the dam which may, in turn, have caused the deficiency in the young with no change in the ability of the placenta to transfer Cu.

Triethylenetetramine (TETA) is another chelating agent used in treating Wilson's disease, particularly when the patient has an adverse reaction to DP. In contrast to DP, which chelates Cu through the sulfhydryl group, TETA binds Cu through its electronegative nitrogens. When TETA was fed to rats,[7] the frequency of resorptions and abnormal fetuses increased with the level fed. Maternal and fetal tissue levels of Cu were decreased by the TETA while there were increases in Zn in the maternal kidney and fetal liver. The results show that the teratogenicity of TETA is probably due to the induction of a Cu deficiency, but the increase in fetal liver Zn may also be a contributing factor. As with DP, it is not known whether any of the effect on the fetus is due to an effect on placental transfer.

IV REPRODUCTION IN COTURNIX

Two studies have been carried out on the effect of EDTA on reproduction in Coturnix. Egg production was not influenced by feeding up to 1.52% EDTA,[8] but was reduced by the 2.28 and 3.04% levels. Fertility was satisfactory except at the highest level, and hatchability was improved above the basal by 0.38 and 0.76% levels, but decreased at higher levels of EDTA. In another study,[9] fertility and hatchability were improved by the addition of 0.5% Na_2·EDTA to a practical diet based upon corn and soybean meal. The effect was not mediated through changes in the Ca, Ba, Fe, Zn, Cu, Ni, or Mn contents of the egg or egg shell thickness. One must assume that the chelating agent somehow improved the mineral environment for the embryo.

REFERENCES

1. **Faber, J. J. and Thornburg, K. L.,** *Placental Physiology,* Raven Press, New York, 1983.
2. **Rubin, M.,** Chelation and iron metabolism, *Am. Feed Manuf. Symp.,* 27, 1968.
3. **Bagley, D. H., Jr., Zapolski, E. J., Rubin, M., and Princiotto, J. V.,** Placental transfer of chelated iron in the guinea pig, *Am. J. Obstet. Gynecol.,* 102, 291, 1968.
4. **Keen, C. L., Saltman, P., and Hurley, L. S.,** Copper nitrilotriacetate: a potent therapeutic agent in the treatment of a genetic disorder of copper metabolism, *Am. J. Clin. Nutr.,* 33, 1789, 1980.
5. **Keen, C. L., Mark-Savage, P., Lönnerdal, B., and Hurley, L. S.,** Teratogenic effects of D-penicillamine in rats: relation to copper deficiency, *Drug-Nutr. Interact.,* 2, 17, 1983.
6. **Mark-Savage, P., Keen, C. L., and Hurley, L. S.,** Reduction by copper supplementation of teratogenic effects of D-penicillamine, *J. Nutr.,* 113, 501, 1983.
7. **Keen, C. L., Cohen, N. L., Lönnerdal, B., and Hurley, L. S.,** Teratogenesis and low copper status resulting from triethylenetetramine in rats, *Proc. Soc. Exp. Biol. Med.,* 173, 598, 1983.
8. **Craig, R. M., Kratzer, F. H., and Vohra, P.,** Growth and reproductive performance of coturnix fed several levels of EDTA, *Poult. Sci.,* 47, 1664, 1968.
9. **Vohra, P., Davis, M. J., and Craig, R. M.,** The improvement of hatchability of Coturnix (Japanese Quail) eggs by EDTA, *Poult. Sci.,* 49, 780, 1970.

Chapter 11

CHELATE TOXICITY

After the recognition that the addition of chelating agents to a diet can improve the utilization of certain trace elements, it became of interest to know the levels at which they can be tolerated in the diet. The work of Suso and Edwards[1] showed improved utilization of only Zn and Mn with the addition of EDTA. There was maximum retention of Zn with the addition of 0.12% (or 4.1 mmol/kg) EDTA to the diet followed by a reduction at lower levels. With Mn, the response was much less, but the maximum retention was with 0.18% EDTA and a level of 0.24% reduced retention (Figure 1). This is compared to the 0.07% level which had been used in most of the experiments of Davis et al.[2] Vohra et al.[3] studied the effect of high levels of EDTA on the deposition of Zn or Cu in the tissue of poults and found no difference in growth with 18.3 or 36.6 mmol of EDTA added per kilogram of feed, equivalent to 0.53 or 1.07%, respectively. There were some definite effects upon the tissue composition with EDTA, in general, counteracting the high levels of Zn, but not Cu.

Solomons et al.[4] studied the effect of the Fe-fortifying agent, NaFe·EDTA upon Zn availability in man by measuring the change in plasma Zn concentration for 6 hr after administering a standard dose of 25 mg of Zn with several test materials. The addition of 15 mg of NaFe·EDTA to the mixture caused no reduction in Zn absorption while 115 and 308 mg definitely caused reduced levels of Zn in the plasma (Figure 2). This raised the question of whether the interference with Zn absorption was due to the EDTA or to the Fe component of the NaFe·EDTA molecule. The administration of 15 mg of Na_2·EDTA with the Zn salt gave no decrease in absorption (Figure 3) and in fact showed a tendency to increase absorption, although the authors found the response nonsignificant. When the level of EDTA was increased to 240 mg, or the equivalent of 308 mg of Na_2Fe·EDTA, there was a flat curve for plasma Zn in one subject, while little effect on Zn absorption of an equivalent level of Fe. The authors concluded that the fortification of sugar with NaFe·EDTA at one part per thousand would not influence Zn nutrition.

The adverse effect of EDTA on Fe absorption has already been mentioned in the section on Fe. Suso and Edwards [1] have shown decreased retention of Fe in whole body counting studies, and Davis et al.[2] found no improvement in Fe utilization with the addition of 0.07% EDTA to a diet of chickens. Chio et al.[5] were able to cause a deficiency of Fe in chickens by the addition of 1.0 or 1.5% Na_2EDTA to a normal diet of practical feedstuffs.

Vohra and Bond[6] fed various levels of EDTA to coturnix chicks in a basal diet of practical feedstuffs. Growth at 2 weeks of age was significantly depressed at EDTA levels of 1% and above, but at 3 weeks of age the differences were not significant, except when 4% of tetrasodium EDTA was used instead of the disodium salt (Table 1). The hematocrit value was greatly decreased by the EDTA, indicating again the interference with Fe utilization. The Zn content of the bones was decreased by the high levels of EDTA while there was an increase in the Zn content of the kidney.

EDTA and NTA are used in certain processed foods to form complexes with certain mineral ions. Since the mineral ions promote the deterioration of food constituents such as unsaturated fatty acids, the chelating agents are effective in improving the stability of the food on storage by reducing the ion concentration in the food. To answer the question of whether these chelating agents would have any deleterious effect on Ca metabolism, Sifri et al.[7] fed NTA at 0.71 or 0.87%, or EDTA at 1.25% along with various levels of Ca to young chicks and young and adult coturnix. NTA or an equimolar level of Na_2·NTA caused a moderate decrease in the weight of the chicks, unrelated to the Ca of the diet. In contrast, feeding Na_2·EDTA caused higher mortality and greater growth depression in chicks fed a

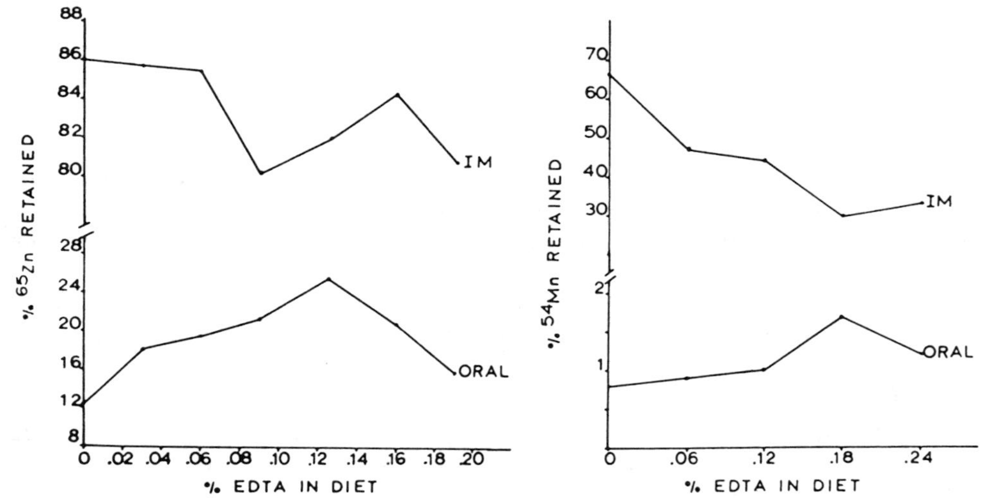

FIGURE 1. The effect of increasing levels of EDTA supplementation on ^{65}Zn and ^{54}Mn absorption. (From Suso, F. A. and Edwards, H. M., *Poultry Sci.*, 47, 1417, 1968. With permission.)

low-Ca diet than a higher Ca level. Plasma Ca levels were significantly increased by NTA in young quail and by Na_2·EDTA in adult quail, whereas they were significantly decreased by Na_2·EDTA in adult quail. Dietary EDTA had little effect upon bone ash while NTA promoted greater tibia mineralization in growing chicks. Chicks appeared to be more sensitive in growth response to dietary chelates than coturnix. Dietary NTA also promoted greater deposition of Zn and Mn in the tibia while EDTA, on the other hand, caused a significant reduction in Zn and Fe.

Citric acid and ascorbic acid are not only natural constituents of foods, but are also often added in food processing. In addition, ascorbic acid is often consumed in amounts far greater than the recommended nutrient allowance because of its reported value in treating the common cold. As a rough approximation, the consumption of 5 g of ascorbic acid per day is equivalent to the addition of 1% to the diet. In experiments with chicks and coturnix, Sifri et al.[8] fed ascorbic acid at various level up to 1% and citric acid at 0.71% at adequate and low levels of Ca. An increase of 1% dietary ascorbic acid had no adverse effect on body weight, tibia ash, or tibia Ca content of adult male chickens over a 224-day period, even at Ca levels as low as 0.026% of the diet. In growing chicks, body weight, plasma Ca and tibia mineral content varied with the Ca level of the diet, but were not influenced by ascorbic acid even at 0.65% of the diet. Citric acid at the same molar level was also ineffective in altering Ca metabolism in growing chickens. There is no evidence in this work that these relatively high levels of ascorbic acid or citric acid have any adverse effect on Ca metabolism because of their chelating properties.

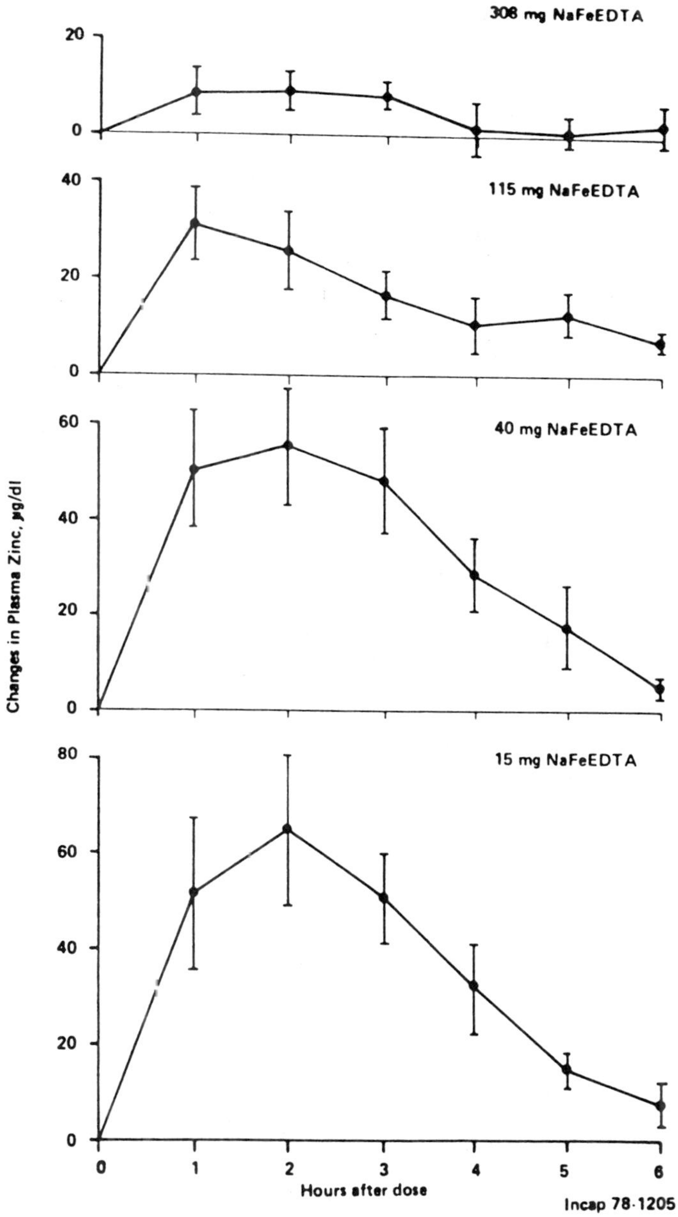

FIGURE 2. Changes in plasma Zn (mean ± SE) over the 6 hr following the administration of 110 mg of $ZnSO_4 \cdot 7H_2O$ mixed into solution with four graded doses of NaFe·EDTA: 15, 40, 115, and 308 mg. Each curve represents five individual tests. (From Solomons, N. W. et al., *J. Nutr.*, 109, 1519, 1979. With permission.)

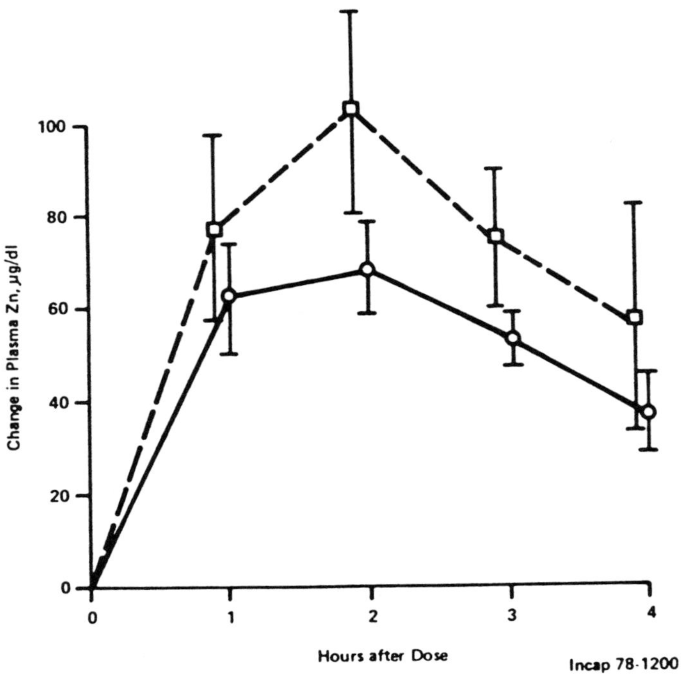

FIGURE 3. Changes in plasma Zn concentration (mean ± SE) following administration of 110 mg of $ZnSO_4 \cdot 7H_2O$ with (dotted line) and without (solid line) 15 mg of disodium EDTA in 100 mℓ of water. Each curve represents the mean of five tests. No differences were significant at any interval. (From Solomons, N. W. et al., *J. Nutr.*, 109, 1519, 1979. With permission.)

Table 1
THE EFFECT OF DIFFERENT LEVELS OF EDTA ON THE GAIN IN BODY WEIGHTS AND BLOOD HEMATOCRIT OF COTURNIX

Experiment number and duration

Treatment	Exp. 1 (28 days)[a]			Exp. 2 (28 days)[b]			Exp. 3 (28 days)[b]	
	Gain (g)[c]		Hematocrit (%)[c]	Gain (g)[c]		Hematocrit (%)[c]	Gain (g)[c]	
Basal diet	83.5 ± 0.8a[d]	(0)[e]	47.0 ± 1.4c[d]	80.7 ± 1.9b[f]	(23)[e]	40.8 ± 0.8b[d]	80.8. ± 0.2b[f]	(11)[e]
+0.5% EDTA	78.1 ± 1.8a	(0)	44.9 ± 1.0bc	75.6 ± 1.5ab	(12)	37.5 ± 0.6b	81.1 ± 2.5b	(5)
+1% EDTA	79.5 ± 2.2a	(0)	40.7 ± 0.7ab	75.2 ± 0.7a	(4)	35.6 ± 1.0ab	76.4 ± 1.4ab	(11)
+2% EDTA	77.5 ± 1.9a	(0)	38.4 ± 1.1a	73.5 ± 1.2a	(8)	31.0 ± 0.4a	77.3 ± 0.2b	(11)
+4% EDTA	75.0 ± 1.4a	(0)	35.5 + 0.5a	64.6 ± 2.1a	(23)	29.2 ± 1.3a	72.4 ± 1.9ab	(24)
+4% EDTA[g]	—		—	—		—	60.2 ± 4.8a	(45)

[a] The coturnix were fed the basal diet for 7 days before feeding the experimental diets.
[b] The experimental diets were fed from the day of hatching.
[c] The values are the means ± standard error.
[d] $p \leq 0.05$.
[e] % Mortality in parentheses.
[f] $p \leq 0.01$.
[g] In this treatment $EDTA\text{-}Na_4$ was used. The other treatments contained $EDTA\text{-}Na_2$.

From Vohra, P. and Bond, D. C., *Poultry Sci.*, 49, 565, 1970. With permission.

REFERENCES

1. **Suso, F. A. and Edwards, H. M.,** Influence of various chelating agents on absorption of ^{60}Co, ^{59}Fe, ^{54}Mn and ^{65}Zn by chickens, *Poult. Sci.*, 47, 1417, 1968.
2. **Davis, P. N., Norris, L. C., and Kratzer, F. H.,** Interference of soy protein with the utilization of trace minerals, *J. Nutr.*, 77, 217, 1962.
3. **Vohra, P., Gottfredson, G. D., and Kratzer, F. H.,** The effects of high levels of dietary EDTA, zinc or copper on the mineral contents of some tissue of turkey poults, *Poult. Sci.*, 47, 1334, 1968.
4. **Solomons, N. W., Jacob, R. A., Pineda, O., and Viteri, F. E.,** Studies on the bioavailability of zinc in man. I. Effects of the Guatamalan diet and of the iron-fortifying agent NaFeEDTA, *J. Nutr.*, 109, 1519, 1979.
5. **Chio, L. F., Bunden, K., Vohra, P., and Kratzer, F. H.,** An abnormal requirement for iron in dystrophic chickens, *Poult. Sci.*, 55, 808, 1976.
6. **Vohra, P. and Bond, D. C.,** The effect of various levels of dietary EDTA on the mineral contents of some tissues of *Coturnix Coturnix* Japonica, *Poult. Sci.*, 49, 565, 1970.
7. **Sifri, M., Lowry, D. C., Kratzer, F. H., and Norris, L. C.,** Effect of NTA and EDTA on calcium metabolism of chickens and coturnix, *J. Nutr.*, 108, 719, 1978.
8. **Sifri, M., Kratzer, F. H., and Norris, L. C.,** Lack of effect of ascorbic and citric acids on calcium metabolism of chickens, *J. Nutr.*, 107, 1484, 1977.

Chapter 12

CHELATES IN METAL DETOXIFICATION AND THERAPEUTICS

I. INTRODUCTION

An excess of any metal can cause toxicity in animals. The ratio between requirement and toxicity may be a small or a large number. A number of the toxic metals coordinate strongly with certain ligands in specific tissue. Some accumulate in the brain (Alzheimer's disease), liver, and kidney and others in the bones. For example, Al tends to concentrate in brain, Pb in bones, and Cd in kidney and testes. The amount that concentrates in the target tissues depends upon the rate of intake and the elapsed time between intake and analysis.

A good number of the metal contaminants from the environment are heavy metals or metalloids with no special biological function and form soft Lewis acids. They often tend to displace the native metal from metalloenzymes by binding more strongly with S-containing ligands than the displaced metal ion. The activity of metalloenzymes is inhibited as a result. The released metal is excreted in the urine. The net result can be changes in the animal behavior as by As, Cd, Pb, Hg, CH_3-Hg, Th, Sn, and V; induction of cancer as by Cr, Ni, As, Cd, and Be; itai-itai-bayo (a bone disease in Japan) by Cd; anemia, changes in membrane permeability, inhibition of protein synthesis and of oxidative phosphorylation.[1] The toxic effects of Fe include mucosal irritation, giddiness, diarrhea, and cardiac collapse.

The absorption of metals is enhanced if they are in a methyl form as compared to their native state. The methylation may be a result of enzymatic or nonenzymatic pathways. An example is the Minimata disaster in Japan caused by the consumption of fish contaminated with methylmercury produced by microbial methylation of Hg(II) in the river bed. Methylmercury is lipophilic and crosses the blood-brain barrier, which cannot be crossed by Hg(II). The accumulation of methylmercury in the brain causes ataxia, speech slurring, deafness, blindness, coma, and death in humans. Se tends to reduce the toxicity due to methylmercury. Methylation of the metals is also observed for As, Pb, Th, Cr, Se, S, Pt, and Pd. Not all of the methylmetal derivatives, however, are more toxic than the original metal.

In the body, the metal may be transferred successively from a ligand with lower to one with a higher binding constant until it becomes concentrated in the target organ from where its removal becomes very difficult. The initial contaminant might be responsible for acute toxicity, but the prolonged exposure results in chronic toxicity. The time interval between the initial exposure and the appearance of toxic symptoms may be quite long. Unless steps are taken for detoxification at an earlier stage, the organism may not survive. Chelating agents are used for detoxification therapy for removal of excessive levels of the metals from the target organs.

The chelating agents compete with binding sites in the biological systems for complexing the metals producing binary or ternary water-soluble complexes. These complexes contain the metal, the therapeutic chelating agent, and another possible ligand. Some of these complexes are excreted through the kidney to urine, and the others are excreted from liver to feces by way of bile. If the filtration load on kidney is heavy, it may get damaged and result in a reduction (oliguria) or cessation (anuria) of urine flow. Hemodialysis or peripheral dialysis is desirable in such cases and kidney function may return to normal.[2]

The following criteria are useful in the selection of a chelating agent for detoxification therapy:[2]

1. The intraperitoneal (i.p.) LD_{50} should be >400 mg/kg. Some data on LD_{50} values are given in Table 1.

Table 1
LD_{50} VALUES OF SOME CHELATING AGENTS[3]

Chelating agent	LD^{50} (mg/kg)	Conditions	Ref.
BAL	105	Rat, i.v.	4
Dimethylglyoxime	250	Rat, oral	4
Deferrioxamine	329	Rat, i.v.	7
D-Penicillamine	334	Mouse, i.p.	5
Na_4·EDTA	380	Mouse, i.p.	4
Ethylenediamine	427	Mouse, s.c.	4
Salicylic acid	891	Rat, oral	4
Acetylacetone	1000	Rat, oral	4
N-Acetyl-D,L-penicillamine	1000	Rat, i.p.	5
2,3-Dimercaptopropane-sulfonic acid	2000	Mouse, s.c.	6
Na_2Ca·EDTA	3800	Rat, i.p.	4
Triethylenetetramine	4340	Rat, oral	4
2,3-Dimercaptosuccinic acid	5000	Mouse, oral	8

a. 4.
b. 5.
c. 6.
d. 7.
e. 8.

2. It should form a very stable complex with the metal ion to be removed.
3. The complex should be water soluble.
4. It should be selective in forming complexes.
5. The complex should be excreted without causing any serious damage to the kidney.
6. The chelating agent should not be metabolized to any significant extent in the body.

A prolonged administration of any nonmetabolizable chelating agent can cause toxicity by depletion of the target metal as well as other essential metals. A metabolizable chelating agent may interfere with the metabolic processes in the body. D-Penicillamine (DP) and triethylenetriamine (TETA) are used as therapeutic agents in Wilson's disease, a genetic disorder characterized by high tissue Cu levels. When administered to rats, both of these chelates caused teratogenic effects. However, these effects were different from those of a Cu deficiency. DP caused a reduction of not only Cu, but also of Zn content, TETA decreased Cu but increased Zn content in the liver. Fe content was elevated.[9] A dietary supplementation with Cu reduced the teratogenic effects of DP, but not of TETA.[10]

It is expected that the chelating agent used for detoxification would have a stronger binding potential for the metal than the tissue ligand with which the metal was complexed in the body. However, no direct correlation between the stability constant of a metal-chelating agent and the ability of the chelating agent to preferentially remove the metal and detoxify the body has been established.

II. METALLOTHIONEINS

Metallothionein was the name given to a protein isolated from equine kidney cortex which contained 5.9% Cd, 2.2% Zn, 0.45% Fe, 0.18% Cu, 14.9% N, and 8.5% S because it was rich in metal and S.[11] Metallothioneins are rich in cysteinyl residues, and may bind one metal more than another. The number of research papers in the field of metallothioneins is extremely large.[12-14]

Metallothionein has been isolated from a wide range of tissues including liver, kidneys, and intestinal mucosa of a number of animals and several cell lines cultured on media containing Zn and Cd. It is also present in fish, marine invertebrates like mussels and crabs, yeasts, and neurospora. Vallee[15] found that metallothioneins could be identified according to the following criteria:

1. The metal content, usually Cd, Zn, or Cu, of these proteins was high.
2. The cysteine (or potential –SH) content was very similar and about 30 to 35% of the total amino acids.
3. The ratio of metal ions bound to –SH groups was 1:3.
4. Generally, 7 g atoms of the metal (Me^{2+}) were present per 20 to 21 –SH groups of the metalloprotein.
5. Aromatic amino acids, histidine, and disulfides were usually absent. A single methionine was present in the mammalian proteins.
6. The protein absorption band at 250 nm was missing.
7. The absorption band was dependent upon the attached metal in the mercaptide form. Cd-thionein absorbed radiation at 250 nm.
8. Their molecular weights ranged between 6000 to 7000.
9. Gel exclusion chromatography was insufficient to recognize a new metallothionein.
10. Metal ions did induce formation of metallothionein.

The function of metallothioneins is not clearly understood, but they are present even in the fetal tissues. A number of functions have been proposed such as homeostasis, transport, specific enzymic activities and catalysis, storage, immune phenomenon, embryonic development, mitogens, cellular differentiation, and detoxification.

The molecular weight of metallothionein is about 6500 and the metal content is about 7 to 10%; cysteine comprises 20 out of the 61 amino acid residues. The values for stability constants for the Cd- and Zn-proteins are around 10^{-15} and 10^{-11}, respectively.[14]

In vitro, metallothioneins may bind heavy metals, but only Cd, Zn, Cu, and Hg are the metals associated with them in vivo. Human liver contains only Zn-thionein.[16]

A pretreatment with a small dose of the contaminant metal such as Cd has been proposed to produce a tolerance to its acute toxicity. The induction of metallothionein by Cd might be responsible for this acquired tolerance.[17] A pretreatment to Pb also helps in later acute Pb toxicity, but Pb does not bind to metallothionein, implying that the mechanism of this Pb tolerance needs a different explanation.

> The following metals have the ability of metallothionein induction: Cd^{2+}, Cu^{2+}, Hg^{2+}, Zn^{2+}, Ag^{+}, Cr^{3+}, Cr^{6+}, Mn^{2+}; In^{3+}, Ni^{2+}, Pb^{2+}, Sn^{4+}, Co^{2+}, Bi^{3+}, Fe^{2+}. The following are incorporated with metallothionein: Cd^{2+}, Cu^{2+}, Hg^{2+}, Zn^{2+}, Ag^{2+}, Sn^{4+}, Au^{+}, Bi^{3+}. If a metal which has the ability to induce Zn-thionein is used as pretreatment, and a metal which can replace Zn in Zn-thionein is used as challenge, the toxicity of challenge metal may be mitigated. However, zinc toxicity is not mitigated with pretreatment with cadmium or copper, because the affinity of cadmium and copper for metallothionein is more powerful than that of zinc. But . . . copper toxicity was not mitigated by zinc pretreatment, although the affinity of copper for metallothionein is more powerful than that of zinc ... This is contradictory to the concept described above.[18]

The exact function of metallothioneins is not clear in detoxification.

III. CHELATION THERAPY

The contaminating metal might enter the body through lungs, skin, or by ingestion. The metal-complexes with a low charge but lipophilic groups cross the membranes very easily. The chelating agent used in decontamination should also be lipophilic/lipophobic to enter the membranes and bind the contaminating metal more strongly than the ligands in the body.

A completely lipophilic chelating agent has a poor tendency to be excreted through the kidneys with the result that the metal-chelate complex may spread to all tissues with drastic results. For excretion, lipophobic chelates are more useful. To accomplish this, the nature of the chelating agent must change from lipophilic to lipophobic, or a lipophobic agent may be needed. The number of ionizable groups on the chelating agent must be more than the charges on the metal contaminant. The lipophobic chelating agent may not cross the membrane, remains in blood and slowly disturbs the equilibrium distribution of the contaminant and eliminates it through the kidneys at a slower rate.

The most commonly detoxifying chelating agents in a decreasing order of softness are as follows: BAL (S donors), DP (N, S, O donors), CaEDTA (N, O donors). The following chelating agents are used in detoxification:

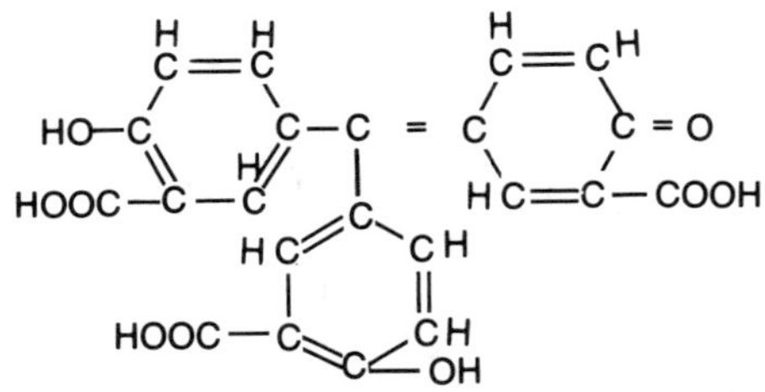

Aurinetricarboxylic acid or Aluminon

It is a lipophilic chelate, its complexes beryllium and the complex is relatively nontoxic. It also complexes Al.

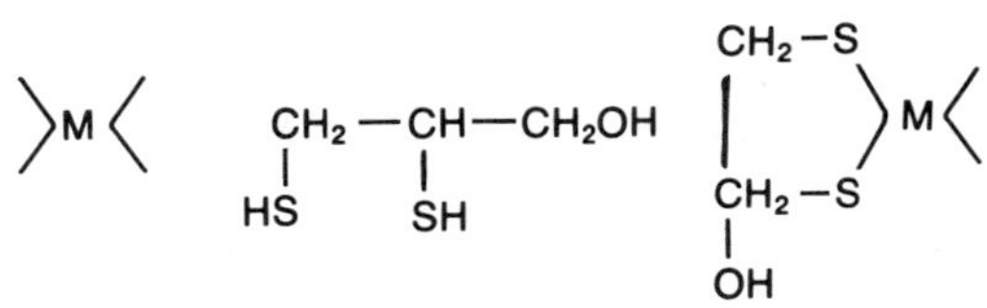

2,3-D-Dimercaptopropanol-1 or British Anti Lewsite (BAL)

It is less effective against Pb, but is most useful against Hg, As, Sb, Cu, Au, Cd, Ti, Tl, and Bi. It may make Hg more soluble and increase its level in brain and should be used with caution. It is used in low concentrations and eliminates the soluble metal complexes through the urine. BAL is also soluble in lipids and can penetrate membranes which may not be permeable to other chelating agents.

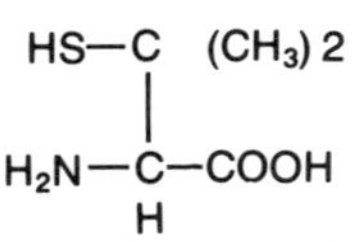

D-Penicillamine (DP)

It is commonly used to remove Cu, as in patients with Wilson's disease. The presence of halide is required for complexation with Cu.

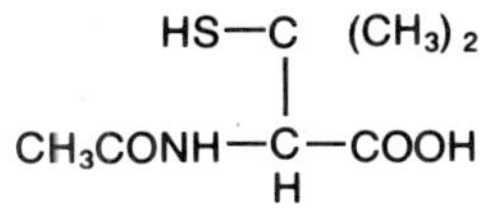

N-Acetyl-D,L-penicillamine (NAPA)

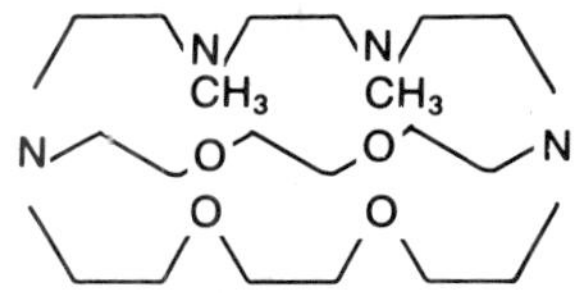

Cryptand

These compounds have high selectivity for Cd^{2+}, which may be removed without affecting Zn^{2+} and Ca^{2+}. Cryptand 2.2.2 is used for the removal of Pb, Sr, and Ra without affecting Ca. It can dissolve $BaSO_4$ in water.

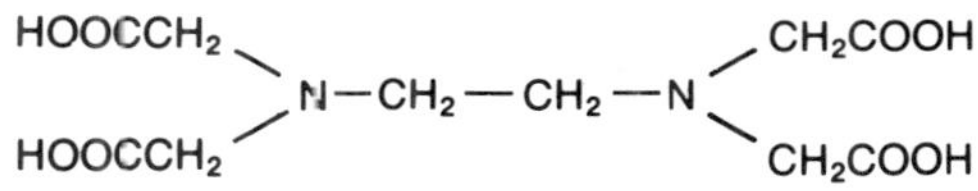

EDTA, Na_2Ca-EDTA, and Ca-EDTA

Injections of EDTA in saline solution can reduce the Ca concentration in the serum to cause tetany and death. For this reason, Ca-EDTA is preferred. Pb detoxification is commonly done with Na_2Ca-EDTA. Na_2Ca-EDTA, rather than EDTA, is used to prevent loss of skeletal Ca. It may be given orally or intravenously. As it is a lipophobic compound, the treatment is very slow. Zn-EDTA is also used because Zn complexes have a higher stability constant than Ca, and it does not deplete Ca from the body.

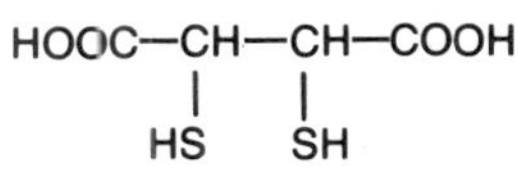

2,3-Dimercaptosuccinic acid (DMSA)

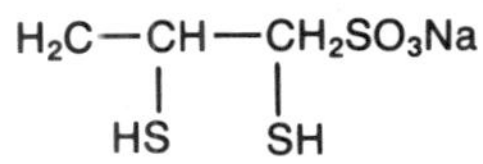

Unithiol or Sodium 2,3-dimercaptopropanesulfonate (DMPS)

It is less toxic than BAL. It is useful in detoxification of Hg, As, Sb, Au, Pb, and Cd. It forms more water-soluble complexes than BAL. It is used to remove Hg from brain. Its Sb complex is used in the treatment of schistosomiasis.[2]

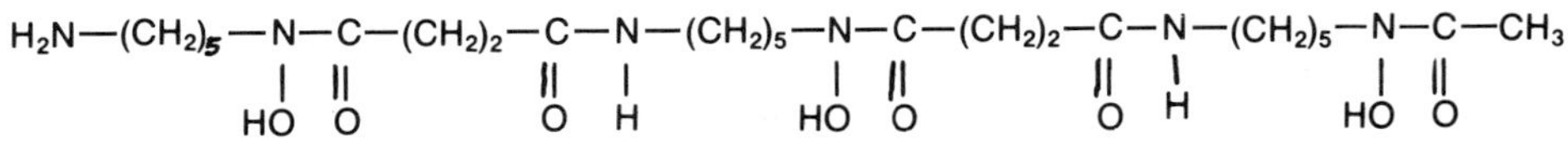

Desferrioxamine

It is used to remove the excess of Fe and is a polyhydroxamic acid. It binds Fe in some microorganisms. The apoprotein of a metalloprotein may be used to remove the excess of that metal.

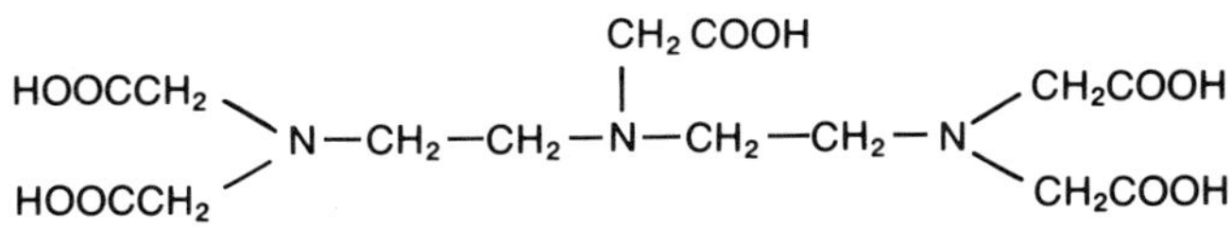

Diethylenetriaminepentaacetic acid (DTPA)

Stability constants of DTPA with metals are higher than of EDTA. Ca and Zn complexes are used in therapy. It is more lipophilic than EDTA and is used against Pu contamination.

IV. OTHER USEFUL CHELATING AGENTS

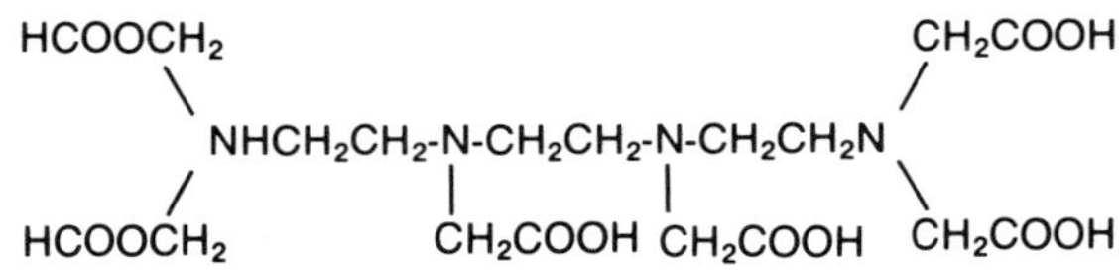

Triethylenetetraminehexaacetic acid (TTHA)

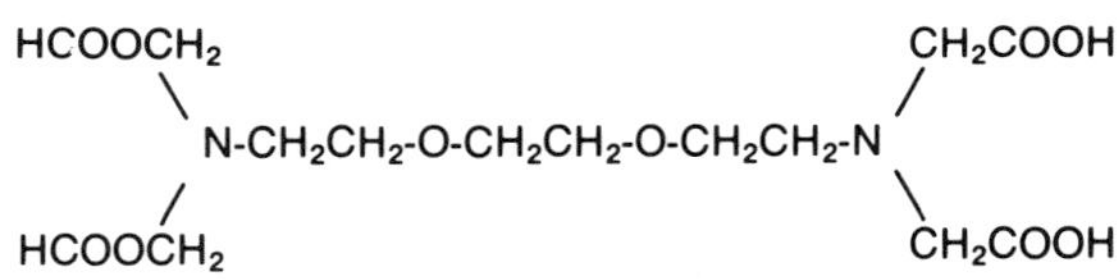

Ethyleneglycol-*bis*-(b-aminoethyl ether) *N,N,N ', N '*-tetraacetic acid or (EGTA)

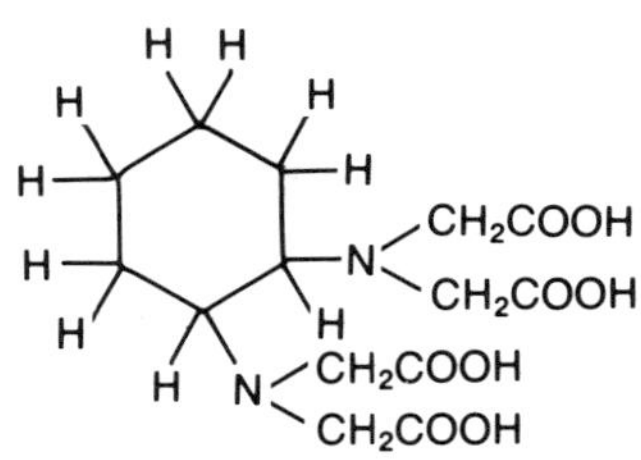

Cyclohexanediaminetetracetic acid or (CDTA)

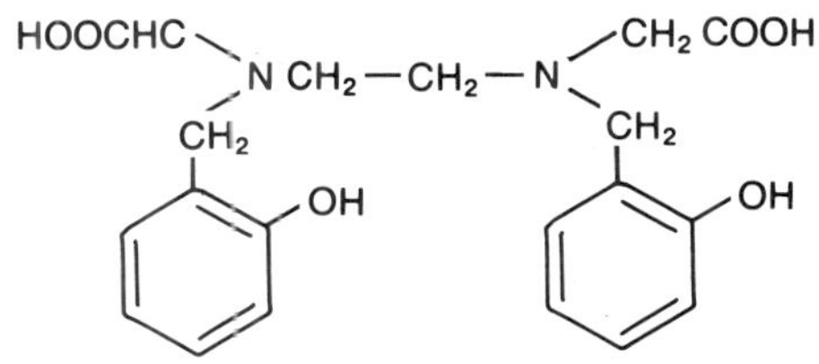

Ethylenebis-*N*,*N* '-(2-*o*-hydroxyphenyl)glycine or (EDDHA)

A. Mixed Chelates

The use of a single chelating agent in detoxification has sometimes increased toxic symptoms. The examples are[19]

- Arsenic by BAL
- Lead by EDTA
- Lead tetraethyl by BAL
- Thallium by BAL
- Mercury by BAL or EDTA
- Cd by BAL, cysteine or $CaNa_2EDTA$
- Beryllium by citrate

For this reason, mixed chelates with proper lipophilic/lipophobic properties are preferred. An example is a mixture of EDTA – BAL.

B. Physiological Function of Chelating Agents

The physiological role of chelating agents may be in maintaining homeostasis.[20]

1. Chelating agents may regulate metal ion concentrations in ionic and chelated form, especially in the low range.
2. Chelation alters oxidation-reduction potential of metal and its subsequent affinity for the ligand. The location of the chelate may shift as a result of charge reduction, increased lipophility, and membrane permeability.
3. Chelates may modify the stereochemical rearrangement of the metabolite, exposing or masking the active centers and thus, influence its participation in competing reactions. The enzyme reactions could be activated or inhibited.
4. Metal ions may act as bridges between chelating organic molecules.
5. Metal-chelates may be involved in electron transport.

C. Drug Function via Chelation

The following drugs appear to function via chelation:

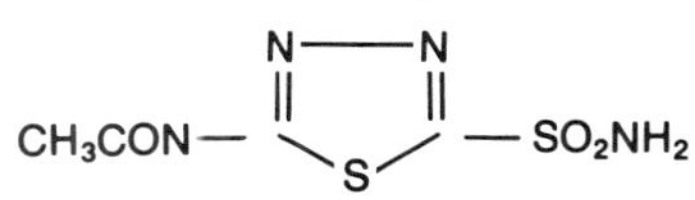

Acetazolamide (carbonic anhydrase inhibitor)

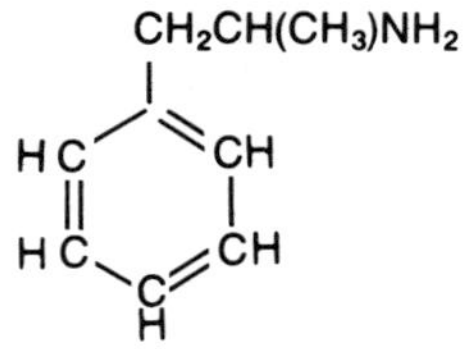

Amphetamine

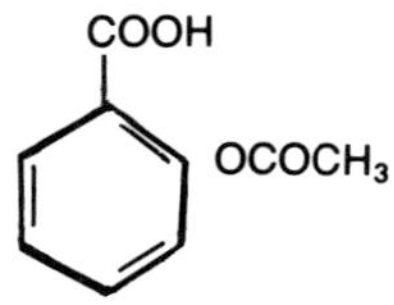

Aspirin

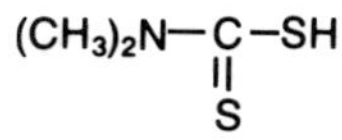

D.M.D.S.

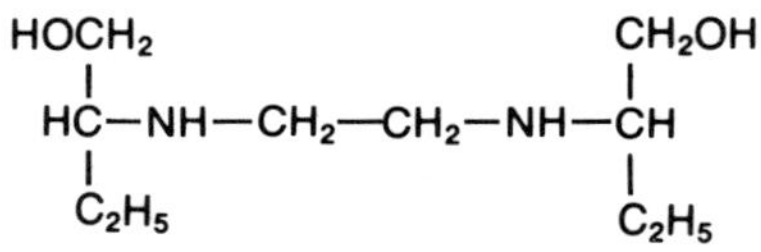

Ethambutol

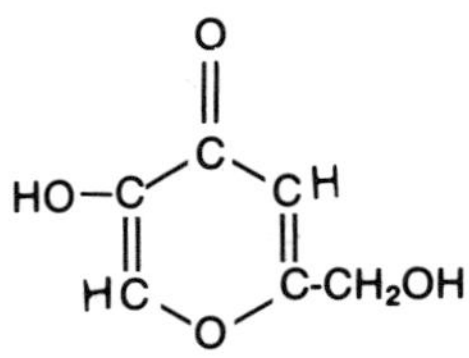

Kojic acid

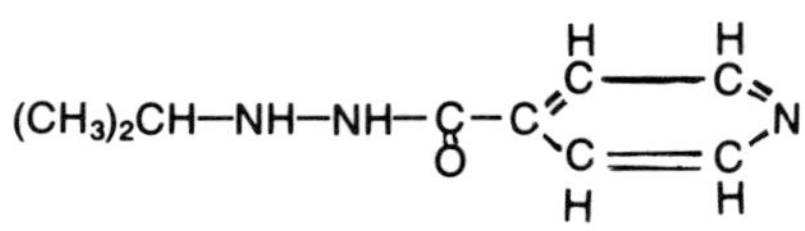

Isoproniazid

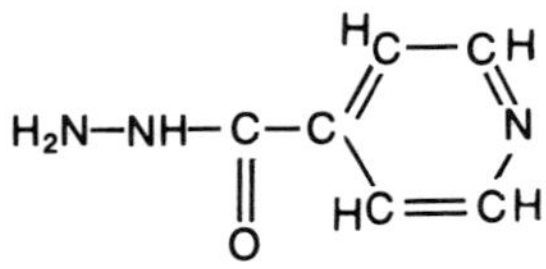

Isonicotinic acid hydrazide

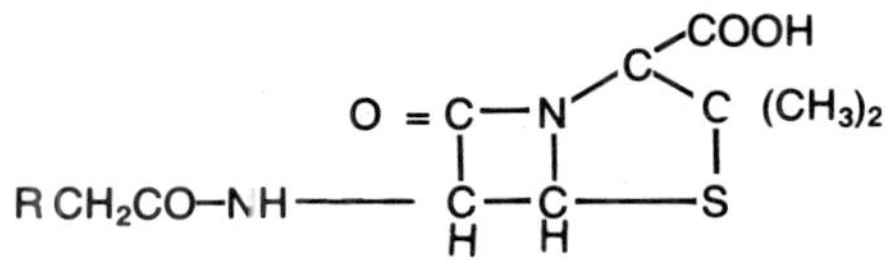

Penicillin

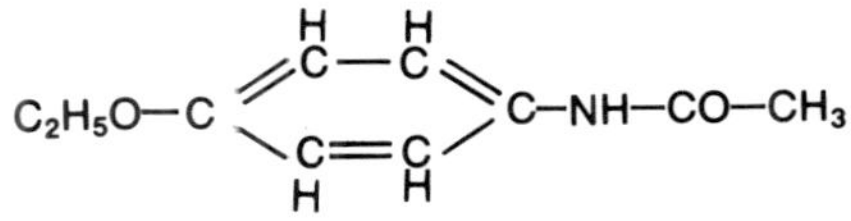

Phenacetin

$C_6H_5-CH_2-NH-CO-CH_2-CH_2-NH-NH-CO-C_5H_4N$

Nialamide

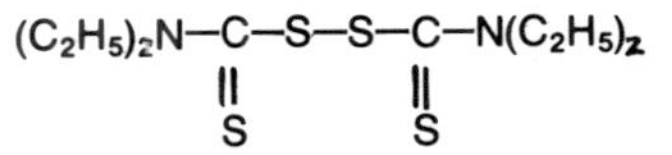

Disulfiram

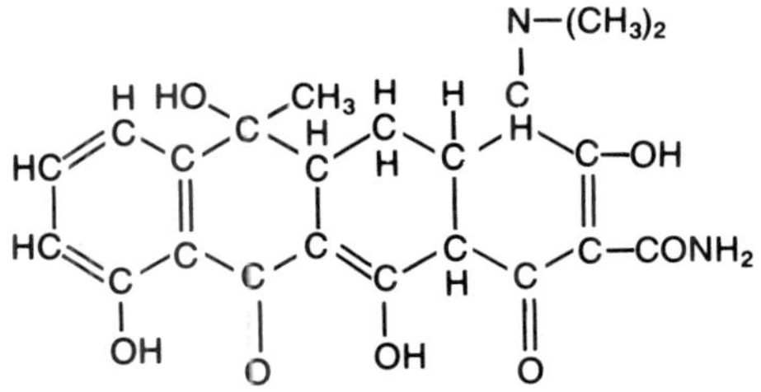

Tetracyclines

$H_2N\ SC\ \ NH-N = CH-(C_6H_4)-NH\ CO\ CH_3$

Thiacetazone

V. DETOXIFICATION OF SOME SPECIFIC METALS

Mercury — Hg(II) has a short half-life in the human body and its removal is difficult after a short critical period. Hg coordinates preferentially with sulfhydryl groups rather than N or O. Chelates containing sulhydryl groups are commonly used for removal of Hg(II) and organomercury compounds. BAL is one of the most commonly used chelates for Hg removal as soluble complexes through the kidney. Without treatment with chelating agents, survival may be only 5%.

Arsenic; Antimony — As(III) and Sb(III) resemble Hg for preferential coordination with sulfhydryl groups. The therapeutic chelates for removal of As(III) and Sb(III) as water-soluble complexes through the kidney are BAL, DMPS, and DMSA.

Manganese — Mn has a poor affinity for sulfhydryl groups, unlike Hg, Cd, and Pb. By forming complexes of low stability, these chelates may alter the distribution of Mn in the body and increase its toxicity. 1,2-Cyclohexylene diaminetetraacetic acid (CDTA), *N*-(2-hydroxyethyl)-ethylenediaminetriacetic acid (HEDTA), and nitriloacetic acid (NTA) were more effective at a lower dose in preventing mortality due to $MnCl_2$ than diethylenetriamine pentaacetic acid (DTPA). Ethylenediamine-*N,N'-bis(o*-hydroxyphenyl) acetic acid (EDDHA) and ethylenediamine-*N,N'*-diacetic acid (EDDA) were less effective or ineffective. Diethylthiocarbamate sodium salt (DDC), triethylenetriamine (TETA), and isonicotinylhydrazine (INH) were slightly effective but the dosage was high. Dimercaptosuccinic acid (DMS), D-penicillamine (DP), and acetyl D-penicillamine (Ac-DP) were ineffective even at a dose of 0.92 mmol/kg for mice.[21]

Cadmium — Metallothionine concentration increased within 2 hr of intravenous (i.v.) Cd administration to mice and reached a maximum within 8 hr. DTPA, EDTA, and DMSA were very effective in detoxification of Cd if administered immediately after exposure to Cd. The efficacy decreased if the chelates were administered after 4 hr.[22] A combination of DTPA and BAL at a level of 50 mg/kg for rats has been used for detoxification against chronic Cd poisoning with minimal kidney damage.[23]

Iron — Ethylenediamine-*N,N'-bis(o*-hydroxyphenylglycine) (EHPG) might be more effective than desferrioxamine (DF) in detoxification of Fe in rats, but its safety still needs to be confirmed.[24]

Lead — The commonly employed chelates had been $CaNa_2$·EDTA, BAL, $CaNa_3$·DTPA, and DPA, and were used singly. It is more desirable to use mixed chelating agents such as EDTA + BAL.

REFERENCES

1. **Huges, M. N.,** *The Inorganic Chemistry of Biological Processes,* John Wiley & Sons, New York, 1972.
2. **Jones, M. K.,** Therapeutic chelating agents, in *Metal Ions in Biological Systems,* Vol. 16, Sigel, H., Ed., Marcel Dekker, New York, 1983, 47.
3. **Jones, M. K. and Basinger, M. A.,** Chelate therapy for type b metal ion poisoning, in *Inorganic Chemistry in Biology and Medicine,* Martell, A. E., Ed., ACS Symp. Ser. 140, American Chemical Society, Washington, D.C., 1980, 335.
4. **HEW,** Toxic Substances, U.S. Department of Health, Education and Welfare, Rockville, Md., 1974.
5. **Aposhian, H. V. and Aposhian, M. M.,** *N*- Acetyl-DL-penicillamine, a new oral protective agent against the lethal effects of mercury chloride, *Pharmacol. Exp. Ther.,* 126, 131, 1959.
6. **Klimova, L. K.,** Pharmacology of the new antidote unithiol, *Farmakol. Toksikol.,* 21, 53, 1958.
7. **Goldenthal, E. I.,** Compilation of LD_{50} values in newborn and adult animals, *Toxicol. Appl. Pharmacol.,* 18, 185, 1971.
8. **Stohler, H. R. and Frey, J. R.,** Chemotherapy of experimental Schistosoma mansoni. Influence of dimercaptosuccinic acid on toxicity and antischistosomal activity of sodium antimony dimercaptosuccinate and other antimony compounds in mice, *Ann. Trop. Med. Parasitol.,* 58, 431, 1964.
9. **Keen, C. L., Mark-Savage, P., Lonnerdal, B., and Hurley, L. S.,** Tertaogenic effects of D-penicillamine in rats: relation to copper deficiency, *Drug-Nutr. Interact.,* 2, 17, 1983.
10. **Keen, C. L., Cohen, N. L., Hurley, L. S., and Lonnerdal, B.,** Molecular localization of copper and inc in the rat fetal liver in dietary and drug-induced copper deficiency, *Biochem. Biophys. Res. Commun.,* 118, 697, 1984.
11. **Kagi, J. H. R. and Vallee, B. L.,** Metallothionein: a cadmium and zinc-containing protein from equine renal cortex. II. Physiochemical properties, *J. Biol. Chem.,* 236, 2435, 1961.
12. **Kagi, H. R. and Nordberg, M., Eds.,** *Metallothionein,* Birkhauser-Verlag, Basel, 1979.
13. **Foulkes, E. C., Ed.,** Biological roles of metallothionein, in *Developments in Toxicology and Environmental Sciences,* Vol. 9, Elsvier/North-Holland, Amsterdam, 1982.

14. **Bremner, I.,** The nature and function of metallothionein, in *Trace Element Metabolism in Man and Animals,* Gawthorne, J. M., Howell, J. M., and White, C. L., Eds., Springer-Verlag, Amsterdam, 1982, 637.
15. **Vallee, B. L.,** Metallothionein: historical review and perspectives, in *Metallothionein,* Kagi, H. R. and Nordberg, M., Eds., Birkhauser-Verlag, Basel, 1979, 19.
16. **Buhler, R. H. O. and Kagi, J. H. R.,** Human hepatic metallothioneins, *FEBS Lett.,* 39, 229, 1974.
17. **Suzuki, Y. and Yoshikawa, H.,** Role of metallothionein in the liver in protection against cadmium toxicity, *Ind. Health,* 12, 141, 1974.
18. **Yoshikawa, H. and Ohta, H.,** Interaction of metals and metallothionein, in *Biological Roles of Metallothionein,* Foulkes, E. C., Ed., Elsevier/North-Holland, Amsterdam, 1982, 11.
19. **Schubert, J.,** Heavy metals — toxicity and environmental pollution, in *Metal Ions in Biological Systems, Studies of Some Biochemical and Environmental Problems,* Dhar, S. K., Ed., Plenum Press, New York, 1973, 239.
20. **Lindenbaum, A.,** A survey of naturally occurring chelating ligands, in *Metal Ions in Biological Systems, Studies of Some Biochemical and Environmental Problems,* Dhar, S. K., Ed., Plenum Press, New York, 1973, 67.
21. **Tandon, S. K. and Khandelwal, S.,** Chelation in metal intoxication. XII. Antidotal efficacy of chelating agents on acute toxicity of manganese, *Arch. Toxicol.,* 50, 19, 1982.
22. **Cantilena, L. R., Jr. and Klaassen, C. D.,** Decreased effectiveness of chelation therapy with time after acute cadmium poisoning, *Toxicol. Appl. Pharmacol.,* 63, 173, 1982.
23. **Cherian, M. G. and Rodgers, K.,** Chelation of cadmium from metallothionein *in vivo* and its excretion in rats repeatedly injected with cadmium chloride, *J. Pharmacol. Exp. Ther.,* 222, 699, 1982.
24. **Hershko, C., Grady, R. W., and Link, G.,** Evaluation of iron-chelating agents in an *in vivo* system: potential usefulness of EHPG, a powerful iron-chelating agent, *Br. J. Haematol.,* 51, 251, 1982.

Chapter 13

REMOVAL OF METALS FROM DIETARY INGREDIENTS

In the preparation of experimental diets for mineral research, it is frequently difficult to obtain a combination of feedstuffs which will be sufficiently low in the mineral to be studied and still be adequate nutritionally. It is usually the protein-rich ingredients which pose the greatest problem, since it is necessary to have the diet adequate in amino acids, and protein sources such as animal or oilseed are usually sources of minerals in various amounts. Diets based upon amino acid mixtures avoid this problem, but they are relatively expensive and, in some instances, unsuitable for the specific study.

The classical diet used for Fe-deficiency studies is based upon dried skim milk as a source of protein. In addition to the protein, however, a diet of 60% dried skim milk contains approximately 30% lactose, which is known to influence the availability of certain minerals such as Fe, Ca, and V. In order to avoid the use of dried skim milk, Davis et al.[1] attempted to use isolated soybean protein in the basal diet for Fe-deficiency studies. The isolated soybean protein contained 61 mg of Fe per kilogram, which was not reduced by washing with water at the isoelectric point of the protein (Table 1). When the isolated soybean protein sample was washed with Na_2·EDTA, however, the Fe content was reduced to approximately half the original level. The procedure involved suspending the soybean protein in tap water at the rate of 454 g per 3.8 ℓ of water. the slurry was heated to 50°C with steam and adjusted to 4.3 (the isoelectric point of the protein) with sodium hydroxide. The Na_2·EDTA was added at the level of 0.5% of the amount of protein and stirred for 30 min. The slurry was then allowed to settle, and the supernatant was syphoned off. After the treatments, the protein was suspended in distilled water, heated, stirred, and allowed to settle. The washing with water was repeated five times. As much water as possible was removed by a press and the soybean protein was dried in an oven at 50°C.

The basal chick diet prepared with the EDTA-treated isolated soybean protein caused reduced growth, lowered hemoglobin and packed cell volume, feather depigmentation, and poorer survival than the diet with untreated protein or the treated protein with the addition of Fe (Table 2). The Fe from the treated isolated soybean protein was less available for the chick than the Fe from the untreated isolated soybean protein. EDTA added to the diet also reduced the availability of Fe, but the EDTA was presumably removed from the treated protein and the poorer availability of the residual Fe was related to the treated protein itself and not to the presence of EDTA.

In addition to the reduction of Fe in isolated soybean protein by the treatment with EDTA, other cations were also reduced (Table 1). Ca, Mg, Mn, Cu, Mo, and Zn were reduced to approximately 10, 4, 25, 40, 30, and 4% of their original levels by the treatment. The use of Na_5·ETPA was about as effective as Na_2·EDTA in lowering the cation content. The P content of the isolated soybean protein was unaltered either by water extraction or EDTA treatment. From the reduction in cation content, it appears that the EDTA treatment procedure would be most useful in preparing diets for studies on any of the cations with possibly the greatest use for Ca, Mg, and Zn.

Isolated soybean protein treated with EDTA was used by Hurley and Swenerton[2] to produce a Zn-deficient diet for rat reproduction studies. The basal ration contained 0 ppm of Zn as determined by X-ray fluorescence analysis, with an error in the method of 2 ppm. Reproduction was not possible with the extreme Zn deficiency. The females showed severe disruption of the estrous cycles and in most cases no mating took place. Thus, it was necessary to use a less deficient diet based upon untreated isolated soybean protein to permit reproduction to occur. The EDTA-treated isolated soybean protein was used successfully in other studies on rats[3] to observe the effects of an extreme deficiency. The signs of the deficiency

Table 1
AVERAGE TRACE MINERAL CONTENT OF UNTREATED AND TREATED SOYBEAN PROTEIN AND OF THE OTHER NONMINERAL INGREDIENTS USED IN THE BASAL DIETS, DRY-MATTER BASIS

	Moisture	mg/kg							
	(%)	Ca	Mg	P	Mn	Cu	Fe	Mo	Zn
Untreated soybean protein, av A and B[a]	8.8	296	442	8530	16	43	61	1.0	51
Water-treated soybean protein, A	6.1	35	28	8763	4	50	68	0.6	32
Na_2·EDTA-Treated soybean protein, 0.5%, 2 times, av A and B	6.8	24	13	8400	6	23	29	0.3	2
Na_2·EDTA-Treated soybean protein, 0.5%, 3 times, A	6.7	48	19	8850	3	16	29	0.4	0
Na_2·EDTA-treated soybean protein, 0.5%, 4 times, A	8.3	29	15	8837	4	7	43	0.3	2
Na_2·EDTA-treated soybean protein, 1.0%, 3 times, A	7.0	65	12	9490	1	16	26	0.3	1
Na_5·ETPA[b]-treated soybean protein, 1.3%, 4 times, A	6.6	37	28	7915	1	20	31	0.1	0
Untreated soybean protein, C[c]	6.1	—	—	8548	8	24	117	7.0	31
Na_2·EDTA-Treated soybean protein, 0.5%, 2 times, C	7.5	43	213	8417	3	14	81	4.0	1
Na_2·EDTA-Treated soybean protein, 0.5%, 4 times, D[d]	9.0	146	28	8854	9	9	48	0	5
Dried skim milk	12.9	9631	308	6817	2	6	28	0	40
Cornstarch	12.5	369	73	448	6	3	12	0	0
Cellulose	0.2	190	108	380	9	4	172	0	4
Soybean oil	0.3	70	4	380	0	1	20	0	0
Chick starter mash	10.8	15,700	1790	10,550	268	16	115	9.8	107

[a] Samples A, B, and C and D are different lots of isolated soybean protein.
[b] Versenex-80, Dow Chemical Company, Midland, Mich.
[c] Sample C was used in experiments 1 and 2.
[d] Sample D was used in experiment 3.

From Davis, D. N. et al., *J. Nutr.*, 78, 445, 1962. With permission.

included extreme retardation of growth, immature hair or alopecia, and dermal lesions. The symptoms were completely reversed by Zn supplementation.

A low Ca diet based upon 14.5% blood fibrin was used in a study of the Ca requirement of adult male chickens.[4] Although the diet contained 0.03% Ca, there was a failure to obtain striking evidence of a Ca deficiency, even during a 14-week feeding period. It was found that the major source of Ca in the diet was blood fibrin, which contained 0.188% Ca. The EDTA treatment method used for the soybean protein was modified to treat the blood fibrin. The fibrin was suspended in deionized water containing Na_4·EDTA, adjusted to pH 11 with sodium hydroxide, and stirred for 1 hr. The pH was then adjusted to 5.2, the isoelectric point of the fibrin, and the fibrin was allowed to settle. The fibrin was treated again with Na_4·EDTA and washed well with deionized water to remove the EDTA.[5] The Ca in the blood fibrin was reduced from 0.188 to 0.008%, or about 4% of the amount in the original sample. This reduced the Ca in the complete diet from 0.03 to 0.0035%. With this low level of Ca it was possible to show a need for Ca in the adult male chicken, although bone mineral is retained very tenaciously at low levels of dietary Ca.

The use of EDTA to lower the mineral content of dietary ingredients has been very useful

Table 2
EFFECT ON GROWTH, HEMOGLOBIN CONTENT, PACKED CELL VOLUME, PIGMENTATION, AND MORTALITY OF CHICKS FED Na_2·EDTA-TREATED SOYBEAN PROTEIN DIETS CONTAINING GRADED AMOUNTS OF Fe AND Cu

Treatment	Body weight[a] (g)	Hemoglobin (g/100 mℓ blood)	PCV[b] (%)	Depigmentation score	Survivors/group
Untreated protein basal diet	353	7.6	23.3	0.0	18/20
+ 80 mg Fe/kg + 6.75 mg Cu/kg	374	9.0	25.6	0.0	19/20
Treated protein basal diet	220	5.0	19.0	0.6	8/20
No Fe + 2.25 mg Cu/kg	257	4.7	16.9	0.6	9/20
No Fe + 6.75 mg Cu/kg	256	5.0	18.5	0.8	9/20
+ 5 mg Fe + 6.75 mg Cu/kg	288	5.1	19.5	0.5	11/20
+ 10 mg Fe + 6.75 mg Cu/kg	273	5.2	18.8	0.5	15/20
+ 20 mg Fe + 6.75 mg Cu/kg	303	5.6	19.5	0.2	15/20
+ 40 mg Fe + 6.75 mg Cu/kg	307	7.2	22.2	0.1	13/20
+ 80 mg Fe + 6.75 mg Cu/kg	366	8.2	23.9	0.1	20/20

[a] Average of duplicate groups.
[b] Indicates packed cell volume.

From Davis, P. N. et al., *J. Nutr.*, 78, 445, 1962. With permission.

for Fe, Zn, and Ca studies. From the reduced mineral content of the treated isolated soybean protein, it is probable that the method could be useful in developing deficient diets for the study of several other minerals as well.

REFERENCES

1. **Davis, P. N., Norris, L. C., and Kratzer, F. H.,** Iron deficiency studies in chicks using treated isolated soybean protein diets, *J. Nutr.*, 78, 445, 1962.
2. **Hurley, L. S. and Swenerton, H.,** Congenital malformations resulting from zinc deficiency in rats, *Proc. Soc. Exp. Biol. Med.*, 123, 692, 1966.
3. **Swenerton, H. and Hurley, L. S.,** Severe zinc deficiency in male and female rats, *J. Nutr.*, 95, 8, 1968.
4. **Norris, L. C., Kratzer, F. H., Lin, H. J., Hellewell, A. B., and Beljan, J. R.,** Effect of quantity of dietary calcium on maintenance of bone integrity in mature White Leghorn male chickens, *J. Nutr.*, 102, 1085, 1972.
5. **Lin, H. J., Norris, L. C., and Kratzer, F. H.,** Extraction of calcium from blood fibrin with a solution of tetrasodiumethylenediaminetetraacetate (Na_4EDTA), *Poult. Sci.*, 52, 540, 1973.

Chapter 14

SUMMARY

Metal chelates are complexes in which the metal atom is held through more than one point of attachment. A chelating agent which yields a soluble metal complex is a sequestering agent. Chelating agents vary in the strength with which they bind metals and in the specificity for binding various metals. The properties of the metal in bound form are far different from the metal in the free state. Both chelating and sequestering agents are found in foods and significantly influence the availability and use of minerals present in and consumed with the food.

The stability constant of a chelating agent for a metal is a measure of the affinity of the metal to complex with the ligand or the chelate. It is greatest when five- or six-membered rings are formed with the ligand. The stability constant increases with the number of rings in the molecule and the basicity of the ligand. It is greatest for more highly charged small molecules.

The animal body contains a host of chelating agents capable of binding metals. Many of the metal chelates, e.g., metalloenzymes, are important for life processes. Other metals may be associated with chelates as in the skeletal system, or with metal-binding proteins in transport systems. In some cases, the metal may be a contaminant which may have no function or may interfere with the function of an essential metal which it has replaced. The affinity of enzymes for metals varies greatly since in some enzymes they are very labile while in others they are bound very strongly.

Foods contain a large number of ligands such as proteins, peptides, amino acids, carbohydrates, lipids, and ionic molecules. Some of these ligands form insoluble complexes from which the minerals may be unavailable to the animal while others may act as sequestering agents and make minerals more available.

Many plant seeds contain phosphates combined with inositol, or phytate, which may be further combined with metals such as Ca, Mg, or K. It may be located in different sites in different seeds, such as in the germ of corn, in the bran of wheat and rice, and throughout the kernel in oilseeds. While cereal and oilseed phytates appear to be quite soluble in water, complexes of phytic acid, protein, Ca, and other minerals may be sparingly soluble in the GI tract. This complex formation may reduce the availability of minerals such as Zn, Mn, Cu, Mg, and Ca, either by the binding of the metal in the complex itself, or its coprecipitation with the complex. Most foods which contain phytic acid also contain fiber. There has been considerable controversy over whether the reduced availability of Zn and Fe in foods is due to the phytate or fiber. It is probable that in many cases, both of these components contribute to the poor availability of minerals. Under some conditions, phosphates alone can also reduce trace mineral availability.

Oxalic acid is present in some foods and forms insoluble complexes with Mg and Ca in the GI tract to reduce their availability. Clays, which are consumed in some cultures, have metal-binding properties and can be important in reducing the availability of trace elements which are marginally adequate in diets.

While phytate, fiber, oxalate, and clays reduce mineral availability, other dietary constituents, such as histidine, cysteine, ascorbic acid, citric acid, fructose, lactose, and certain proteins, are effective in improving their utilization. This has been shown in a variety of ways, such as increasing growth in mineral-deficient animals, increasing the uptake of radioactive minerals, and preventing deficiency symptoms in animals.

In addition to responses to known constituents of foods, there are well-conducted experiments which have shown greater physiological responses to the addition of materials such as liver extract or distillery byproducts in mineral-deficient diets than in adequate ones. This suggests that there may be factors (chelates) present which improve the availability of the marginal level of the mineral in the diet. It is possible that some of the unidentified factor response which has been observed frequently in nutritional studies is due to improved mineral balance caused by chelates, rather than to a component with metabolic significance.

In the GI tract there is thus a heterogenous mixture of ligands from many sources competing for the trace elements. The availability of a mineral depends to a large extent upon the ligands with which the particular mineral becomes associated.

Synthetic chelating agents can improve or impair the utilization of dietary minerals, depending upon their specific properties. The effectiveness of chelating agents to improve Zn availability varies with the published stability constants for the chelates, reaching a maximum at about 14, after which it declines. The significance of this observation is not clear, however, because the published stability constants were determined in a nonphysiological pH range and values determined at the physiological pH are not of the same order.

There are many sites for the control of mineral utilization, depending upon the mineral which is being considered. The most probable mechanism by which chelating agents improve the utilization of minerals depends upon their sequestering ability or the ability to compete with other ligands in the GI tract to form soluble complexes with the mineral. In some cases the mineral may be taken to the brush border where it is released for absorption by the epithelial cell while in most cases there is evidence that the chelating agent is absorbed with the metal which it binds. The metal may be released to other ligands which have greater affinity for it than the chelating agent with which it was absorbed. In many cases, the chelating agent may be metabolized, thereby releasing the metal for its physiological function. In some cases there may be multiple functions of chelating agents, as for example, the action of ascorbic acid as a reducing agent for Fe, as well as a specific function in absorption.

Nutritionists like to have precise values for nutritional requirements which can be compared to the nutrient content of a diet to predict the adequacy of a nutrient in the diet. There are several factors, such as phytic acid and fiber, in food which can reduce the availability of minerals, as well as other factors, such as some amino acids and ascorbic acid, which can improve the availability of minerals. For this reason, it is impossible to give an exact value for a requirement which would apply to all diets. This is not to say that requirement values are not useful in evaluating diets, but one must be aware that there are several dietary constituents which can greatly influence the requirement values and the availability of the minerals present in a diet.

Chelating agents have been shown to influence the availability of minerals to the fetus. Some of the effect may be due to increasing the level of the mineral in the dam's circulation but beyond that, some chelating agents specifically improve placental transport to the fetus. There may be a place for dietary chelates in influencing the mineral content of the fetus with an inherited defect in mineral metabolism.

In addition to improving the utilization of a mineral, chelating agents can also influence the toxicity of a mineral element. This does not follow a clear-cut pattern since with some minerals the toxicity is reduced by the chelate and with others the toxicity is increased. The use of chelating agents to remove toxic minerals from the body is well established. In contrast to the use of chelating agents to improve mineral absorption where it is desirable that the chelate be metabolized, in detoxification use it is desirable that the chelate combines with the toxic metal in the body to be excreted as such without being metabolized.

The dietary requirement for a mineral may be greatly reduced by the addition of a chelating agent to a diet. A practical question arises about whether one should correct a marginal mineral deficiency in a diet by adding the mineral, a chelate, or a chelated mineral. While it might be possible to use the chelate, or chelated mineral, it is a matter of relative costs that will influence the decision. In most cases, the cost of the mineral supplement itself is less than that of a chelate or chelated mineral.

INDEX

A

B

D

G

H

I

K

L

M

N

O

S

T

U

V

W

X

Z